13th Anglo-French Physical Acoustics Conference

(AFPAC 2014)

Journal of Physics: Conference Series Volume 581

Croydon, United Kingdom
15 – 17 January 2014

ISBN: 978-1-63439-933-3
ISSN: 1742-6588

Printed from e-media with permission by:

Curran Associates, Inc.
57 Morehouse Lane
Red Hook, NY 12571

proceedings.com

Some format issues inherent in the e-media version may also appear in this print version.

Printed by Curran Associates, Inc. (2015)

Institute of Physics
Dirac House, Temple Back
Bristol BS1 6BE UK

Phone: 44 1 17 929 7481
Fax: 44 1 17 920 0979

techtracking@iop.org

Additional copies of this publication are available from:

Curran Associates, Inc.
57 Morehouse Lane
Red Hook, NY 12571 USA
Phone: 845-758-0400
Fax: 845-758-2634
Email: curran@proceedings.com
Web: www.proceedings.com

TABLE OF CONTENTS

Journal of Physics: Conference Series **581** (2015) 011001 doi:10.1088/1742-6596/581/1/011001

13th Anglo-French Physical Acoustics Conference (AFPAC2014)

The 13th Anglo-French Physical Acoustics Conference (AFPAC) was held at Selsdon Park Hotel, Croydon near London, United Kingdom, on 15-17 January 2014. The venue was an excellent location to exchange ideas, regardless whether this happened in the conference room, over lunch at the drinks reception in the conservatory, in the oak panelled bar after the conference dinner or in the local pub next door. Over 45 papers were presented at the conference. There were over 60 delegates from institutions covering four countries.

The invited speakers from the French side shared their knowledge about the generation of sound from supersonic jets (Prof Christophe Bailly, École Centrale de Lyon) and the application of ultrasonic microscropy in the nuclear industry (Prof Gilles Despaux, Université de Montpellier). The UK invited speakers included Prof Malcolm Povey (University of Leeds), who talked about characterisation of the nucleation of crystals using ultrasound, and Prof Bruce Drinkwater (University of Bristol), who captured the audience by speaking about "ultrasonic lassos" and ultrasonic particle manipulation.

There was a strong representation of laser ultrasonics at the meeting with scientific considerations of problems and applications that range from the macro to the nanoscale. There were also numerous papers on the interaction of elastic and acoustic waves with complex materials and scattering of these waves by materials such as foams or cavitating liquids. Presentations on biomedical applications are increasingly being featured at AFPAC meetings. Talks this year covered topics such as imaging and high-intensity focused ultrasound for therapeutic applications. Finally, there were also several contributions from the field of Non-Destructive Evaluation (NDE) and Structural Health Monitoring (SHM) with talks ranging from the determination of the properties of *in vivo* wood to ultrasonic scattering techniques and tomographic reconstructions to recover the size and shape of defects in pipes and plates.

The UK organising committee was particularly happy to welcome the many French contributors that travelled to Croydon and would like to thank Alain Lhémery and the Société Française d'Acoustique (SFA) for publicising the event in France. We are happy to announce that many of the presented papers will be published in the conference proceedings in Journal of Physics: Conference Series. We would also like to draw the reader's attention to the upcoming 14th AFPAC conference that is scheduled to take place on 14-16 of January 2015 in Fréjus, France.

Pierre Gélat (UCL, United Kingdom), Valerie Pinfield (Loughborough University, United Kingdom), Frederic Cegla (Imperial College, United Kingdom), Nader Saffari (UCL, United Kingdom) and Alain Lhémery (CEA, France)

Peer review statement

All papers published in this volume of *Journal of Physics: Conference Series* have been peer reviewed through processes administered by the proceedings Editors. Reviews were conducted by expert referees to the professional and scientific standards expected of a proceedings journal published by IOP Publishing.

Journal of Physics: Conference Series **581** (2015) 012001 doi:10.1088/1742-6596/581/1/012001

Ultrasonic Histotripsy for Tissue Therapy

K J Pahk[1], D K Dhar[2], M Malago[3] and N Saffari[1]

[1]Department of Mechanical Engineering, University College London, WC1E 7JE, UK
[2]Institute of Liver and Digestive Health, University College London, NW3 2PF, UK
[3]Hepato-pancreatic-biliary and Liver Transplantation Surgery, Royal Free Hospital, University College London, NW3 2PF, UK

E-mail: k.pahk@ucl.ac.uk

Abstract. Hepatocyte transplantation has been considered and investigated as a promising and alternative method to liver transplantation for treating liver-based metabolic disorder in newborns over the past two decades. Although some clinical trials have been conducted and shown clinical benefits and outcomes, it is difficult to deliver and achieve a desired level of integration and transplantation of hepatocytes in the liver parenchyma. To overcome this problem, this work introduces an alternative method to a portal-infused-hepatocyte cell transplantation. To improve the level of engraftment of transplantable hepatocytes, these are injected directly into cavities generated by ultrasonic histotripsy. Histotripsy is an extracorporeal noninvasive technique which has been recently developed using high intensity focused ultrasound (HIFU) for inducing tissue fractionation with no coagulative necrosis. The exact mechanisms for the tissue fractionation are not well understood yet; but the possible mechanisms are thought to be a combination of nonlinear wave propagation effect, explosive bubble growth and ultrasonic atomization. The main objectives of this work are to demonstrate the feasibility of this new cell therapy and evaluate and distinguish between the different types of cavitation activity for either a thermally or a mechanically induced lesion. In the present work, numerical studies on the bubble dynamics (the Gilmore-Akulichev bubble model coupled with the Khokhlov-Zabolotskaya-Kuznetsov equation) and both *ex*- and *in vivo* liver experiments are conducted with histological analysis (haematoxylin and eosin stain). The numerical and the experimental results suggest that (a) the acoustic emissions emitted during the thermal ablation and the histotripsy exposure can be distinguished both numerically and experimentally and (b) the proposed cell therapy may potentially form an effective and safe clinical treatment for replacing and correcting disordered hepatocytes, although the transplantation was not conducted in this work.

1. Introduction

Liver transplantation is the main and ultimate clinical treatment for most patients with inherited liver diseases [1]; however donor shortage is currently the main limitation [2]. Hepatocyte transplantation has been suggested and proposed as an alternative promising method to liver transplantation over the past two decades [3], [4], [5]. Despite the significant clinical benefits of hepatocyte transplantation one of the main limitations is the low level of engraftment of transplanted hepatocytes due to the instant blood mediated inflammatory reaction (IBMIR) and the liver to body weight ratio making it difficult to achieve extensive proliferation of transplanted cells [6], [7]. To overcome such problems, this work introduces an alternative method to a portal-infused-hepatocyte cell transplantation. A proposed method is to produce a cavity inside the liver parenchyma created by ultrasonic histotripsy which acts as a hepatocyte receptor and which facilitates the successful uptake, proliferation and integration of the transplanted cells into the disordered or diseased liver after direct injection into the cavity.

Recently, there has been significant interest in producing tissue fractionation rather than inducing coagulative necrosis using high intensity focused ultrasound (HIFU). This is defined as histotripsy [8]. A lesion produced by histotripsy contains fractionated tissue with no evidence of thermal damage around it [9]. The exact mechanisms for the tissue fractionation are not well understood yet; but the possible mechanisms are thought to be a combination of nonlinear wave propagation effect, explosive bubble growth and ultrasonic atomization [10]. Histotripsy uses high amplitude shock waves to fractionate tissue and has been demonstrated using both micrometre-sized cavitation bubbles (maximum radii of the order of a 100 μm) with microsecond HIFU pulses (cavitation cloud histotripsy) [11] and millimetre-sized boiling bubble for milliseconds of HIFU excitation (boiling histotripsy) [12].

In this paper, a single spherical bubble is modelled using the Gilmore-Akulichev (GA) equation coupled with the Khokhlov-Zabolotskaya-Kuznetsov (KZK) equation to investigate bubble dynamics (including corresponding radiated pressures) excited by both a histotripsy and a thermal ablation waveform. *Ex vivo* experiments are conducted to obtain acoustic emissions from a thermally and a mechanically induced lesion. Finally, an *in vivo* experiment is performed in rat liver to demonstrate the feasibility of the proposed cell therapy mediated by histotripsy.

2. Numerical method

2.1. Acoustic fields-The KZK equation

In HIFU applications, a commonly used equation for simulating an acoustic field is the Khokhlov-Zabolotskaya-Kuznetsov (KZK) parabolic nonlinear wave propagation equation. It is an augmented form of Burgers' equation and accounts for the effects of diffraction, absorption and nonlinearity. The axisymmetric form of the KZK equation in terms of acoustic pressure is given by [13]

$$\frac{\partial^2 p}{\partial z \partial \tau} = \frac{c_0}{2}\nabla_\perp^2 p + \frac{\delta}{2c_0^3}\frac{\partial^3 p}{\partial \tau^3} + \frac{\beta}{2\rho_0 c_0^3}\frac{\partial^2 p^2}{\partial \tau^2} \tag{1}$$

where p is the acoustic pressure, z is the propagation coordinate along the axis of the wave beam, c_0 is the small signal sound speed, ρ_0 is the ambient density, $\beta = 1 + B/2A$ is the coefficient of nonlinearity where B/A is a parameter of the nonlinearity of the medium, δ is the diffusivity of sound, $\tau = t - z/c_0$ is the retarded time variable and $\nabla_\perp^2$ is the transverse Laplacian operator. The sound diffusivity is given by

$$\delta = \frac{1}{\rho_0}\left[\left(\mu_B + \frac{4\mu_s}{3}\right) + \kappa\left(\frac{1}{c_v} - \frac{1}{c_p}\right)\right]$$

where μ_B and μ_s are bulk and shear viscosity, κ is thermal conductivity, c_v and c_p are specific heat capacities at constant volume and at constant pressure respectively.

The HIFU Simulator 1.2 [14] was used to compute acoustic pressures and waveform at the focus as a function of the electrical power applied to a HIFU transducer. This open source software is widely used and validated experimentally by other HIFU studies in cavitation cloud histotripsy [15], cavitation enhanced heating in bovine liver [16], HIFU beams [17], temperature measurements in *ex vivo* porcine muscle [18] and HIFU lesion formation in *ex vivo* swine tissues [19]. The KZK numerical results are illustrated in subsection 3.1.

2.2. *Bubble dynamics: the Gilmore-Akulichev equation*

In this work, the Gilmore-Akulichev bubble model (GA) is used for simulating gas bubble dynamics because it takes into account the compressibility of the fluid and the variation of sound velocity in the liquid (equation of state: $P = A(\rho/\rho_o)^m - B$; $\partial P/\partial\rho = c^2$ where A, B and m are material-dependent constants). The assumptions behind this model make it suitable to study single spherical bubble dynamics subjected to high acoustic pressure amplitudes such as in lithotripter shock wave pulses [20],[21]. The Gilmore-Akulichev equation is a second order differential equation which governs the dynamics of a single bubble subjected to a compressible liquid, and is given by [21]

$$\ddot{R}R\left(1-\frac{\dot{R}}{C}\right)+\frac{3}{2}\left(1-\frac{\dot{R}}{3C}\right)\dot{R}^2=\left(1+\frac{\dot{R}}{C}\right)H+\frac{\dot{R}}{C}\left(1-\frac{\dot{R}}{C}\right)R\frac{dH}{dR} \tag{2}$$

and the time-varying pressure (outward) radiated by the bubble motion in the liquid, $P_r(t)$ is [21],

$$P_r(t)=A\left[\frac{2}{m+1}+\frac{m-1}{m+1}\left(1+\frac{m+1}{rc_0^2}G\right)^{1/2}\right]^{2m/(m-1)}-B \tag{3}$$

where

- R – is the bubble radius;
- $\dot{R}$ – is the velocity of the bubble wall;
- $\ddot{R}$ – is the acceleration of the bubble wall;
- $C=\{c_0^2+(m-1)H\}^{1/2}$ – is the speed of sound at the bubble wall;
- $H=\frac{m}{m-1}\frac{A^{1/m}}{\rho_0}\{[P(R)+B]^{(m-1)/m}-[P_\infty(t)+B]^{(m-1)/m}\}$ – is the enthalpy;
- $P(R)=P_g-\frac{2\sigma}{R}-4\mu\frac{\dot{R}}{R}$ – is the pressure at the bubble wall;
- $P_g=\left(P_0+\frac{2\sigma}{R_0}\right)\frac{n}{n_0}\left(\frac{R_0}{R}\right)^{3\gamma}\left(\frac{R_{0n}}{R_0}\right)^{3(\gamma-1)}$ – is the pressure of gas inside the bubble;
- R_{0n} – is the time-varying equilibrium radius of the bubble;
- R_0 – is the initial bubble radius;
- n – is the instantaneous number of moles of gas in the bubble;
- n_0 – is the initial quantity of moles of gas in the bubble;
- c_0 – is the speed of sound in the liquid;
- P_0 – is the ambient pressure of the surrounding liquid;
- P_∞ – is the pressure at infinity;
- r – is a distance from the centre of the bubble ($r \gg R(t)$ [22], set to 10 cm);
- $G=R(H+\dot{R}^2/2)$ – is an invariant of the bubble motion;
- μ – is the viscosity of the liquid;
- σ – is the surface tension of the liquid;
- γ – is the polytropic index of the gas inside the bubble.

For simplicity, gas-rectified diffusion across the bubble wall is ignored here (i.e., let $n = n_0$ and $R_{0n} = R_0$). To solve the Gilmore-Akulichev equation numerically, it is firstly nondimensionalised according to the following scheme:

$$Mass=\frac{P_0R_0}{(2\pi f)^2};\ Length=R_0 \text{ and } Time=\frac{1}{2\pi f}$$

The nondimensionalised Gilmore-Akulichev equation is then integrated numerically using an explicit Runge-Kutta method (4), (5) routine using *ode45* in MATLAB (MathWorks Inc., R2013a).

During the numerical computation, it is assumed that a single spherical gas bubble is initially at rest, the bubble remains spherical during its oscillation, there is no mass transfer and no heat diffusion at the bubble wall and the bubble pulsates adiabatically. Acoustic fields used in the Gilmore-Akulichev equation are obtained from the KZK simulation. The GA numerical results are shown in subsection 3.2.

Parameters used for both the acoustic fields and the bubble dynamics simulations are shown in table 1 and table 2 respectively.

Table 1. Parameters used in the KZK simulation. A 1.1 MHz (Sonic Concepts, H102) and a 2.0 MHz (Sonic Concepts, H106) HIFU transducer were employed for the *ex vivo* and the *in vivo* experiments respectively. The HIFU focus was set to 5 mm below the surface of the liver.

Symbol	Definition	HIFU transducer	
		1.1 MHz	2.0 MHz
a	Outer radius of HIFU transducer	3.2 cm	3.2 cm
b	Inner radius of HIFU transducer	1 cm	0 cm
d	Focusing depth	6.32 cm	6.32 cm
F_{width}	Focal width	1.37 mm	0.73 mm
F_{depth}	Focal depth	10.21 mm	7 mm
e_{ff}	Efficiency of HIFU transducer	85 %	85 %
G	Linear pressure gain	35.4	72.9
K	Number of harmonics used in the KZK simulation	200	200
		Water	Liver [23]
c_0	Small-signal speed of sound	1481 ms^{-1}	1568 ms^{-1}
ρ_0	Mass density	1000 kgm^{-3}	1060 kgm^{-3}
α	Attenuation at 1 MHz	0.217 dBm^{-1}	52 dBm^{-1}
η	Exponent of attenuation vs. frequency curve	2	1.1
β	Nonlinearity parameter	3.5	6.8

Table 2. Parameters used in the GA equation. The liver is modelled as a highly viscous human blood-like tissue.

Symbol	Definition	Water [24]	Liver
σ	Surface tension	0.073 Nm^{-1}	0.056 Nm^{-1} [25]
μ	Viscosity of fluid	1.046×10^{-3} $kgm^{-1}s^{-1}$	9×10^{-3} $kgm^{-1}s^{-1}$ [25]
γ	Polytropic exponent of the gas	1.4	1.4
P_0	Ambient pressure	1.01×10^5 Pa	1.01×10^5 Pa
m	Material dependent constant	7 at 20 °C	5.527 (human blood) [26]
A	Material dependent constant	300.1 MPa	614.6 MPa (human blood) [26]
B	Material dependent constant	$A - P_0$	$A - P_0$

3. Numerical results

3.1. Acoustic fields

Acoustic fields in the liver were obtained using the KZK simulation as a function of the electrical power applied to a HIFU transducer. The calculated acoustic peak positive (P_+) and peak negative (P_-) pressures used for both thermal ablation and boiling histotripsy are shown in table 3.

Table 3. Acoustic peak positive (P_+) and peak negative (P_-) pressures at the focus. The acoustic pressures used for boiling histotripsy are within the range of required pressure values reported by other HIFU histotripsy studies [27], [28], [29].

	1.1 MHz HIFU transducer		
Electrical power [W]	P_+ [MPa]	P_- [MPa]	Description
60	13.4	-7.5	Thermal ablation (for *ex vivo*) for inducing coagulative necrosis.
350	74.4	-14.4	Boiling histotripsy (for *ex vivo*) for producing tissue fractionation.
	2.0 MHz HIFU transducer		
Electrical power [W]	P_+ [MPa]	P_- [MPa]	Description
200	101	-16.7	Boiling histotripsy (for *in vivo*) for producing tissue fractionation.

3.2. Bubble dynamics

In boiling histotripsy, an explosive bubble growth is one of the possible mechanisms to cause tissue fractionation and that growth is thought to be led by nonlinear wave propagation effect [30]. The effect of the acoustic waveform on the bubble dynamics (i.e., sinusoidal, boiling histotripsy and thermal ablation waveforms) is numerically investigated in the absence of any heat or mass transfer across the bubble wall.

3.2.1. Boiling histotripsy – Nonlinear-shocked wavefront. To study the effect of boiling histotripsy waveform on a gas bubble in a liquid, the bubble dynamics driven by both the histotripsy and the sinusoidal waveforms are compared here. The peak negative pressure of the sinusoidal wave is matched to that of the histotripsy waveform shown in figure 1.

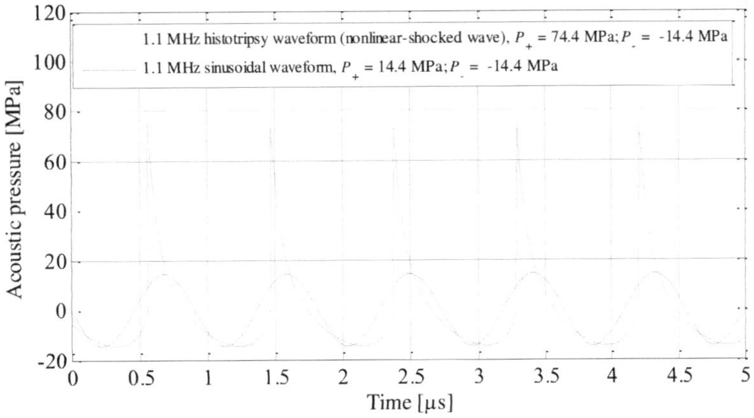

Figure 1. Histotripsy (red) and sinusoidal (blue) waveform with a frequency of 1.1 MHz. Both peak negative pressures are matched to -14.4 MPa. The histotripsy wave (nonlinear-shocked waveform) was obtained from the KZK simulation.

The initial bubble size was chosen as 20 μm [30] as this generated rectified bubble growth after only a few acoustic cycles. In figure 2, the response of the bubble is plotted for both the histotripsy (P_+ = 74.4 MPa and P_- = -14.4 MPa) and the sinusoidal waveform (P_+ = 14.4 MPa and P_- = -14.4 MPa). For the sinusoidal wave excitation, the bubble continues to grow and collapse with time; whereas the nonlinear-shocked waveform leads the bubble to undergo rectified growth within a few acoustic cycles.

These different bubble behaviours can be described by the asymmetry in the compressional and rarefactional pressure phase. In the nonlinear waveform, the positive pressure phase has a shorter duration than the negative pressure part; therefore, the bubble has relatively longer time to undergo expansion rather than contraction; resulting in the explosive growth. As the bubble gets larger over each acoustic cycle, its characteristic time becomes longer and eventually the bubble responds more to the tensile portion [30]. In the sinusoidal excitation case, it can be seen in figure 2 that the bubble behaviour is not periodic. This is likely due to the large acoustic pressure magnitude and the presence of nonlinear resonance (i.e., the ratio of the driving frequency and the linear resonance frequency approaches a rational number) [31], [32].

The rectified bubble behaviour can also be seen with the initial bubble radius of 7 μm, P_+ = 101 MPa and P_- = -16.7 MPa (used in an *in vivo* experiment) at a driving frequency of 2.0 MHz (figure 3).

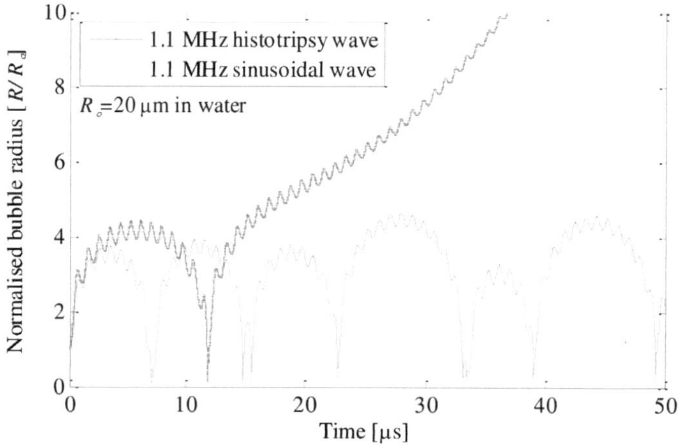

Figure 2. Rectified growth of a 20 μm bubble in water with a driving frequency of 1.1 MHz. The red and blue solid line represent the bubble dynamics under histotripsy (P_+ = 74.4 MPa and P_- = -14.4 MPa) and sinusoidal waveform (P_+ = 14.4 MPa and P_- = -14.4 MPa) respectively.

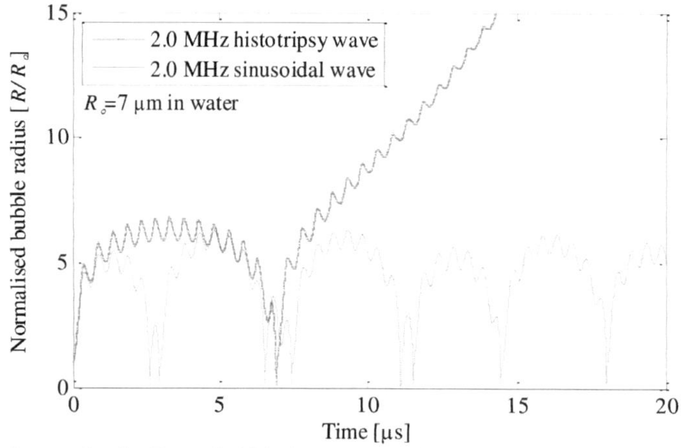

Figure 3. Rectified growth of a 7 μm bubble in water with a driving frequency of 2.0 MHz. The red and blue solid line represent the bubble dynamics under histotripsy (P_+ = 101 MPa and P_- = -16.7 MPa) and sinusoidal waveform (P_+ = 16.7 MPa and P_- = -16.7 MPa) respectively.

Further qualitative observations may be made by adding both the mass and heat transport into the Gilmore-Akulichev equation. Kreider *et al* [30] showed numerically that the rectified bubble growth was observed at a high ambient temperature of 80 °C.

3.2.2. Thermal ablation waveform – a slightly distorted nonlinear wavefront. In contrast to the explosive bubble growth driven by the nonlinear-shocked-waveform shown in figure 2, no rectified bubble growth is observed with the thermal ablation waveform (figure 4b). This is most probably because the degree of nonlinearity (i.e., absence of a shock in the waveform) and the duration of the tensile part (negative pressure phase) are not sufficient to lead to such rectified behaviour. Nevertheless, the bubble growth under the thermal ablation waveform is enhanced compared with that under the sinusoidal wave since it has a slightly longer negative pressure part than that of the sine wave.

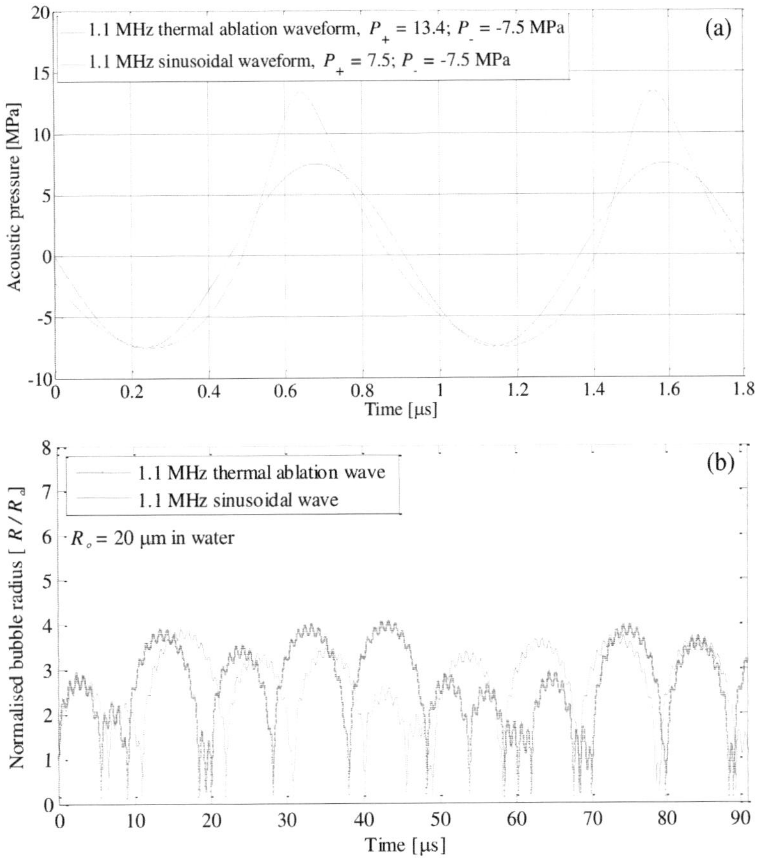

Figure 4. Thermal ablation waveform (P_+ = 13.4 MPa and P_- = -7.5 MPa) and the corresponding bubble dynamics at a frequency of 1.1 MHz. (a) shows the thermal ablation (red) and the sinusoidal waveform (blue). (b) The corresponding bubble dynamics with the thermal ablation and the sinusoidal waveform. A 20 μm bubble is in water. The thermal ablation waveform was obtained from the KZK simulation.

3.2.3. Bubble dynamics in the liver modelled as a highly viscous human blood-like liquid. Forced bubble oscillations in the liver-like-fluid are investigated here instead of using the Keller-Miksis [33] equation which accounts for a weakly compressible liquid. This was coupled with the linear Kelvin-Voigt viscoelastic model proposed by Yang and Church [34]. According to the numerical results shown in figure 5 the rectified growth behaviour also appears under the boiling histotripsy waveform (figure 5b and c), whereas there is no such growth with the thermal ablation excitation (figure 5a). For both the histotripsy and the thermal ablation waveforms, the magnitude of the oscillations of the bubble is always small compared to that in water (i.e., suppressed), because of higher viscous damping in the liver.

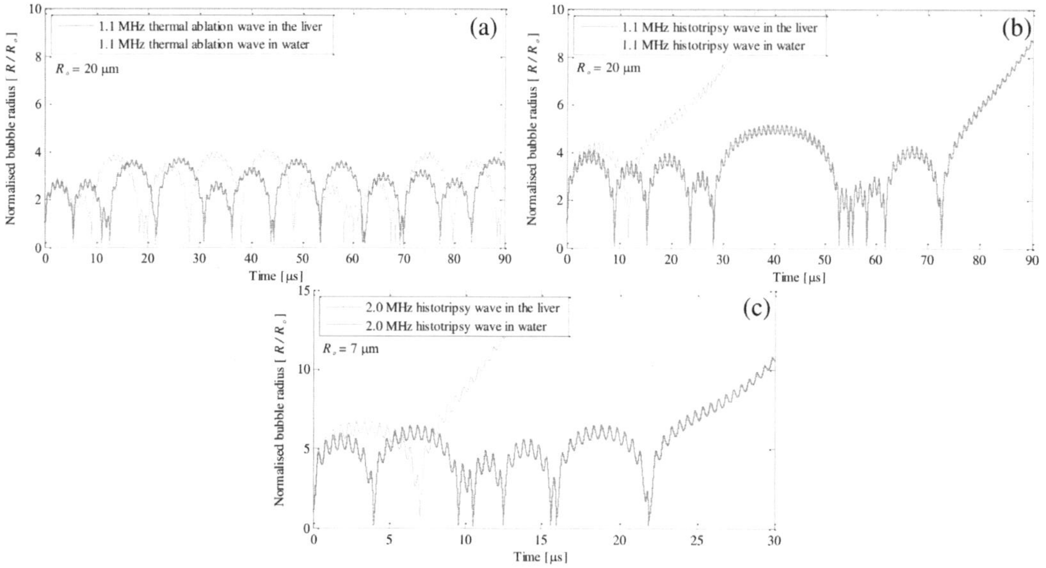

Figure 5. Rectified bubble growth in the liver-like fluid. A 1.1 MHz with (a) P_+ = 13.4 MPa and P_- = -7.5 MPa; (b) P_+ = 74.4 MPa and P_- = -14.4 MPa and (c) a 2.0 MHz with P_+ = 101 and P_- = -16.7 MPa. Blue and red solid lines indicate the bubble dynamics in water and the liver respectively.

3.3. Radiated pressure in the liver

The acoustic pressure radiated by an oscillating bubble driven at a frequency of 1.1 MHz was computed using equation (3) to understand the changes in the emitted acoustic signal with respect to a change in the incident acoustic field (histotripsy and thermal ablation waves) and the initial bubble radius.

In figure 6, the computed frequency spectra of the radiated pressure in the liver surrounding a 2, 5 and 10 μm bubble exposed to both the histotripsy (blue solid line, P_+ = 74.4 MPa; P_- = -14.4 MPa) and the thermal ablation waveform (red solid line, P_+ = 13.4 MPa; P_- = -7.5 MPa) are plotted.

According to figure 6, there are multiple harmonics present (of the form nf_0, n is integer) as well as distinct frequencies at f_0/n (sub-harmonics), nf_0/m (ultra-subharmonics, m is integer and $n<m$), mf_0/p (ultra-harmonics, p is integer and $m>p$) and a broadband component. In addition, these harmonics and the broadband component increase significantly with respect to a reduction of the initial radius. These differences in the frequency spectrum are because of the resonance frequency of a given initial bubble radius and the ratio of the resonance to the excitation frequency (i.e., different characteristic time scales for motion at a given bubble radius). A smaller bubble, for instance, has a higher resonance frequency (i.e., short time-scale) which results in more oscillations in a given period of time and eventually leads the bubble to grow faster and larger compared with a larger bubble with a lower resonance frequency. According to Neppiras [35] and Gaete-Garreton *et al* [36] the appearance of these harmonics as well as the broadband spectrum is generally considered an indicator of strong cavitation.

By comparing the frequency spectrum of the radiated pressure produced by the bubble dynamics under the histotripsy and the thermal ablation excitation over 100 acoustic cycles (figure 6d), it can be observed that (a) multiple harmonics of larger amplitude as well as (b) the elevation of broadband emissions appear with the histotripsy wave excitation. This is likely to be due to the presence of a shock in the histotripsy waveform which contains higher multiple harmonics and pressure amplitudes, resulting in the higher degree of nonlinearity in the radial bubble oscillations and larger amplitude short-duration pressure spikes caused by inertial cavitation (i.e., combined effects of the nonlinear wave propagation and the presence of inertial cavitation).

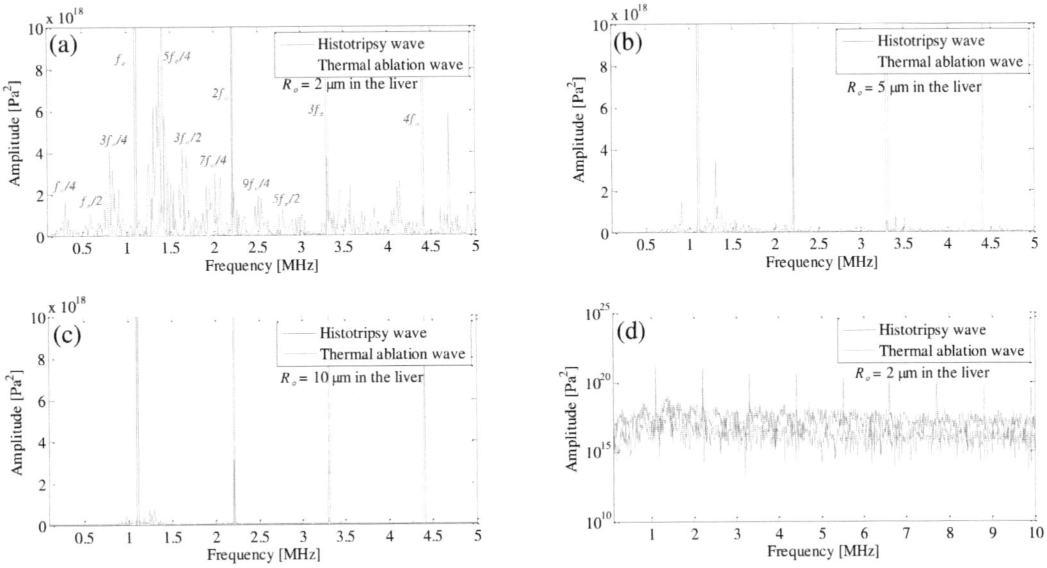

Figure 6. Computed frequency spectra of the radiated pressure from (a) 2, (b) 5 and (c) 10 μm bubble exposed to a 1.1 MHz histotripsy (blue solid line, P_+ = 74.4 MPa; P_- = -14.4 MPa) and thermal ablation wave (red solid line, P_+ = 13.4 MPa; P_- = -7.5 MPa) over 100 acoustic cycles. (d) corresponds to a plot of (a) in a logarithmic scale on the *y*-axis. The multiple harmonics and the broadband component are most probably due to a nonlinear radial bubble oscillation (excited by the nonlinear waveform) and short-duration pressure spikes emitted by collapsing bubble respectively [37].

4. *Ex vivo* experiment

4.1. *Experimental method*

An *ex vivo* experiment was carried out to record acoustic emissions in order to evaluate and distinguish between the different types of cavitation activity for either a thermally or a mechanically induced lesion. A schematic diagram of the *ex vivo* experimental setup is shown in figure 7. The experiment was conducted in a water bath (50×30×22 cm) filled with 16.5 L of distilled water at room temperature (20 °C). The distilled water was then degassed using a water treatment system (Precision Acoustics, WTS). An oxygen meter (VWR, OX 4000H) was used to measure the gas concentration and the experiment was carried out once the concentration reached 20 %.

A 1.1 MHz HIFU source (Sonic Concepts, H102) was used consisting of a focused single element transducer with an aperture size of 64 mm with a central hole of 20 mm in diameter (focal length of 63.2 mm). The transducer was driven by a function generator (Agilent, 33220A) via a RF power amplifier (ENI, 1040L). A computer with waveform generation software (Agilent, Agilent Waveform Builder) was used to design and trigger the function generator for various HIFU pulsing protocols. To measure the level of electrical power supplied to the transducer, a power meter (Sonic Concepts, 22A) was connected between the amplifier and the transducer. *Ex vivo* chicken liver was placed in water at a temperature of 20 °C. During the experiment, a 10 MHz 15-multielement ultrasound imaging probe

(Siemens, Sonoline Versa Plus) was in contact with the tissue sample. The HIFU focus was set to 5 mm below the surface of the liver (see figure 7).

For recording the acoustic emissions produced at the focus in the sample during the HIFU exposure, a 10 MHz single element focused transducer (Olympus, V327) was inserted in the middle of the HIFU source and aligned confocally with the HIFU focus. The focused hydrophone was then linked to an analogue signal filter (Kronh-Hite, 3945). The filter type was set to Butterworth-bandpass (2 to 15 MHz) and it was connected to a digital oscilloscope (LeCroy, HDO 6054).

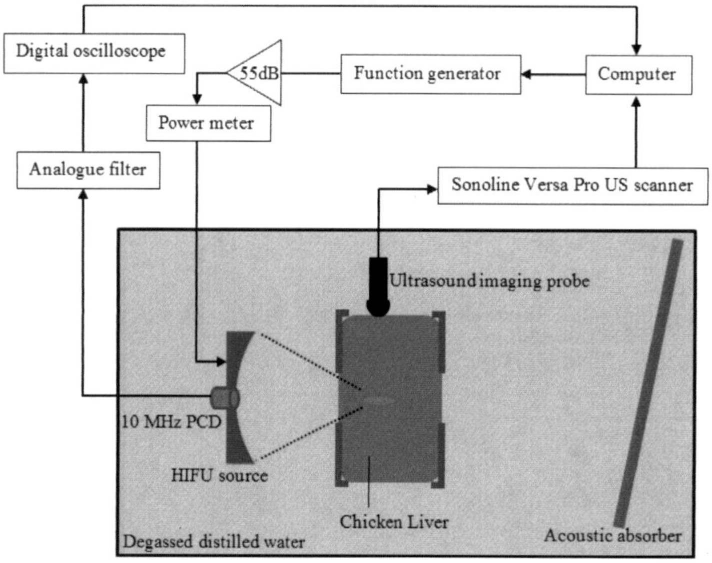

Figure 7. A schematic diagram of the *ex vivo* experimental setup used.

4.1.1. Signal processing for Passive Cavitation Detection (PCD). The sampling frequency of the oscilloscope was set to 125 MHz. The raw signals (filtered through the analogue filter during the experiment) were deconvolved by the transfer function of the focused PCD transducer using a Wienner filter. The deconvolved signals were then filtered again using a 10th order Butterworth highpass filter (2 MHz cut-off) through MATLAB (MathWorks Inc., R2013a). The signals were subsequently analysed in voltage vs time plot, an amplitude vs frequency plot (using a Fast Fourier Transform) and a frequency vs time plot (spectrogram, using a Short Time Fourier Transform).

4.1.2. Experimental protocol-1.1 MHz HIFU transducer. In *ex vivo* experiment, two sets of HIFU exposures were used for inducing a thermally and a mechanically induced lesion. The electrical power supplied to a 1.1 MHz HIFU source was varied between 60 W (P_+ = 13.4 MPa; P_- = -7.5 MPa) and 350 W (P_+ = 74.4 MPa; P_- = -14.4 MPa) for producing a thermally and a mechanically induced lesion respectively. A continuous 5 s HIFU excitation was used for the thermal ablation exposure, whereas the duty cycle, pulse length and pulse repetition frequency for the HIFU histotripsy insonation were chosen according to Wang *et al* [27], Khokhlova *et al* [28] and Canney *et al* [29], shown in table 4.

Table 4. HIFU histotripsy exposure setting used during the *ex vivo* experiment. 10 ms ON time for a pulse.

Duty cycle [%]	Pulse length [ms]	Pulse repetition frequency [Hz]	Number of pulses
1	10	1	50

4.2. Experimental results

4.2.1. 1.1 MHz histotripsy exposure for inducing a cavity. Figure 8 shows the cross sectioned cavity produced inside the *ex vivo* liver and the corresponding ultrasound B-mode images before and after insonation. In general, the shape of a lesion generated by histotripsy is a tadpole like consisting of a "head" and a "tail" [38]; however the formation of the "head" with the absence of the "tail" is observed in figure 8a. This is most probably due to the presence of the blood vessel within the "head" boundary which the fractionated tissue travelled through during the exposure. Examining the tissue morphology around the cavity, there is no evidence of thermal damage at boundary of the cavity such as blanching.

Figure 9 illustrates the corresponding acoustic emissions recorded over period of 10 ms. There are noticeable differences in the acoustic emissions before and after 4.3 ms. In the voltage vs time plot, the amplitude is fairly uniform; it however increases significantly at 4.3 ms (figure 9a). After 4.3 ms, the peak-to-peak voltage amplitude starts to fluctuate violently. The corresponding spectrogram also indicates these changes (figure 9b). The spectrogram contains horizontal stripes from 2.2 to 9.9 MHz (multiple harmonics from second to ninth) before 4.3 ms; however stronger and more noticeable stripes with higher harmonics up to 13.2 MHz (11th harmonic) are suddenly observed at 4.4 ms. These multiple harmonics are likely to be due to the combination of the nonlinear radial bubble oscillations as well as the reflection of the incident nonlinear-shocked acoustic waves from the bubbles filled with gas and vapour due to the large acoustic impedance mismatch at their surface. Because the incident acoustic field was maintained constant during the insonation, these significant changes must be related to both the level of nonlinear radial oscillations and the surface area of the bubbles (assuming the bubble does not affect the incident acoustic field). In this context, it can be conjectured that the significant appearance of a local maximum at 4.3 ms in the voltage vs time plot, as well as in the spectrogram, is indication of the boiling bubble [29]. Because the bubble acts as stronger scatterer (due to a relatively larger surface area) and oscillates with a highly nonlinear radial behaviour, the acoustic emission is, therefore, strong and contains more multiple harmonics.

On the other hand, prior to the onset of the boiling bubble at 4.3 ms, the cavitation activity produces a lower peak-to-peak voltage amplitude and a fairly uniform frequency spectrogram. In the spectrogram, the cavitation activity encompasses multiple harmonics as well as the broadband component. Similarly, these multiple harmonics are due to the combination of the reflection of the HIFU incident wave and the nonlinear radial bubble motions, whereas the broadband component is the result of the short-duration pressure spikes caused by the inertial cavitation [37]. These experimental observations are consistent with Canney *et al* [29].

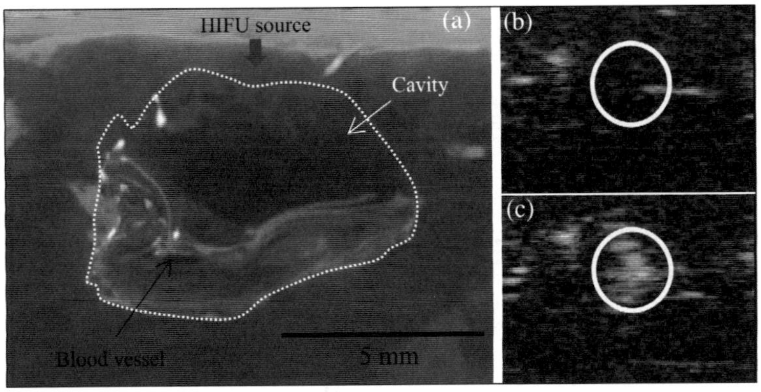

Figure 8. Cavity produced by a 1.1 MHz HIFU transducer. Electrical power of 350 W (P_+ = 74.4 MPa and P_- = -14.4 MPa), duty cycle of 1 %, PRF of 1 Hz, 10 ms pulse length and 50 pulses were used. (a)

Cross sectioned cavity. Ultrasound B-mode images (b) before the HIFU insonation and (c) after one pulse.

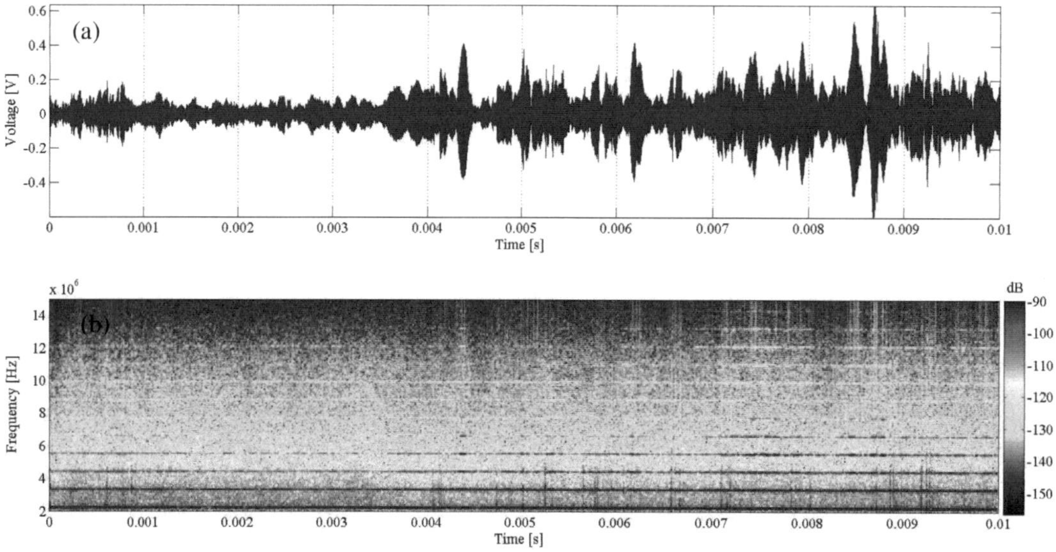

Figure 9. Corresponding acoustic emissions produced by the first HIFU boiling histotripsy pulse in the *ex vivo* tissue. The emission was obtained for 10 ms. (a) illustrates the filtered voltage vs time plot and (b) is the corresponding spectrogram.

The Bioheat transfer (BHT) simulator [14] was used to compute the temperature rise at the focal region in order to confirm the appearance of the boiling bubble (i.e., indication of the focal temperature of 100 °C) in the PCD measurement at 4.3 ms (figure 9). The BHT equation is given by [39]

$$\rho C \frac{\partial T}{\partial t} = k\nabla^2 T + H - wCT \tag{4}$$

where ρ is the tissue density, C is the heat capacity, w is the perfusion rate, k is the thermal conductivity and H is the heat source. The heat source can be obtained from the KZK simulation. Parameters used for the temperature simulation are shown in table 5.

Table 5. Parameters used in the BHT simulation.

Symbol	Definition	Liver [23]
C	Heat capacity	3628 [$Jkg^{-1}K^{-1}$]
k	Thermal conductivity	0.57 [$Wm^{-1}K^{-1}$]
w	Perfusion rate	0 for the *ex vivo* liver [$kgm^{-3}s^{-1}$]
T	Ambient temperature	20 [°C]

Figure 10 shows the temperature rise of the *ex vivo* liver tissue exposed to a 1.1 MHz nonlinear-shocked waves (boiling histotripsy, P_+ = 74.4 MPa and P_- = -14.4 MPa). The computed time to reach a boiling temperature of 100 °C is 4 ms, which is similar to the PCD measurement (figure 9, 4.3 ms) with 7.5 % of the differences between the PCD measurement and the BHT simulation.

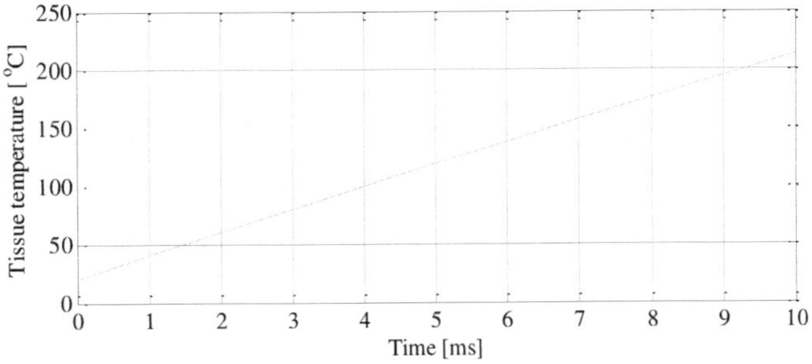

Figure 10. Computed temperature of an *ex vivo* liver tissue exposed to a 1.1 MHz nonlinear-shocked waves (boiling histotripsy, P_+ = 74.4 MPa and P_- = -14.4 MPa) over 10 ms. The time to reach a boiling temperature of 100 °C is 4 ms.

4.2.2. 1.1 MHz thermal ablation exposure for inducing coagulative necrosis. Figure 11 shows the thermally ablated *ex vivo* liver sample and the corresponding ultrasound B-mode images. In contrast to the cavity produced with the boiling histotripsy insonation, there is a well-defined thermal lesion (i.e., coagulative necrosis) which corresponds to the ellipsoidal shape of the HIFU focal region. In addition to the coagulative necrosis, there are three pits (blue arrows) within the lesion and these are the results of inertial cavitation [40].

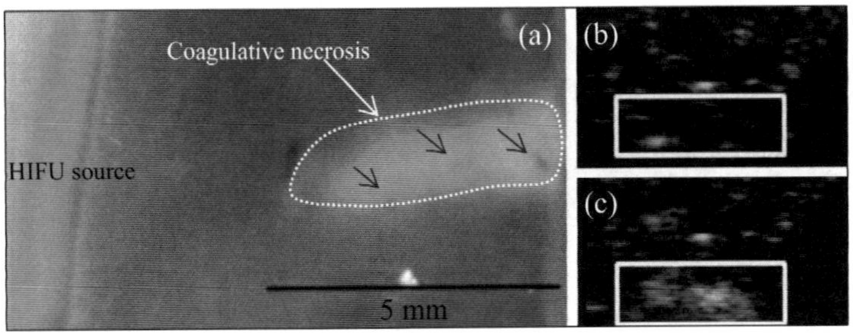

Figure 11. Thermally induced lesion by a 1.1 MHz HIFU transducer. Electrical power of 60 W (P_+ = 13.4 MPa and P_- = -7.5 MPa). A continuous 5 s HIFU insonation was used. (a) Cross sectioned thermal lesion with three pits (blue arrows). Ultrasound B-mode images (b) before and (c) after the exposure.

The corresponding spectrogram of the acoustic emissions obtained over 10 ms (HIFU exposure time was 5 s) is plotted in figure 12. In contrast to the emissions produced by the histotripsy exposure shown in figure 9, there is no indication of the boiling bubble in the spectrogram. This is mainly because the magnitudes of both the peak positive and negative pressures and the shape of the incident acoustic waveform (i.e., absence of a shock) were not sufficient to produce a boiling bubble within 10 ms (i.e., the time to reach a boiling temperature of 100 °C is 2.2 s as shown in figure 13). Although there is no indication of the presence of a boiling bubble, the recorded acoustic emissions suggest that cavitation may be occurring (figure 12), but that this is not due to the large bubble formed by boiling.

Journal of Physics: Conference Series **581** (2015) 012001 doi:10.1088/1742-6596/581/1/012001

Furthermore, at this acoustic pressure, no cavity was observed with the same duty cycle, pulse length, pulse repetition frequency and number of pulses used in the previous histotripsy experiments in subsection 4.2.1. Khokhlova *et al* [28] reported that the purely mechanical damage was only observed when both boiling and shockwaves were present at the focal region.

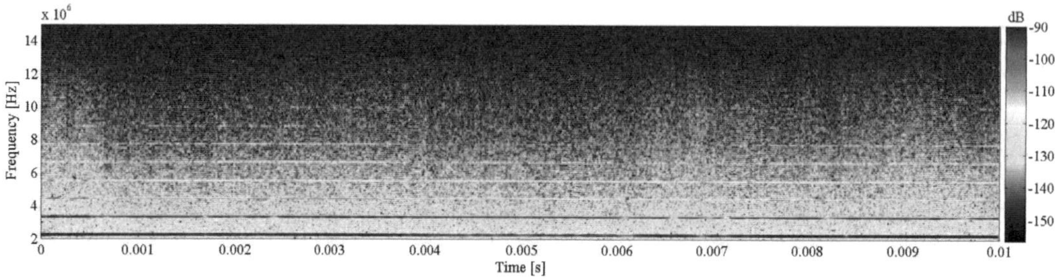

Figure 12. Corresponding spectrogram of the acoustic emissions produced by the HIFU thermal ablation exposure in the *ex vivo* tissue. The emission was obtained only for 10 ms.

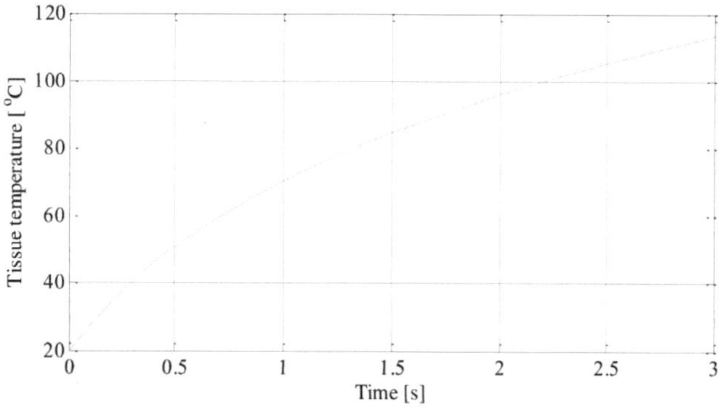

Figure 13. Computed temperature of an *ex vivo* liver tissue exposed to a 1.1 MHz nonlinear waves (thermal ablation, P_+ = 13.4 MPa and P_- = -7.5 MPa). The time to reach a boiling temperature of 100 °C is 2.2 s.

4.3. Ex vivo experimental results vs radiated pressure simulations

In this subsection, experimentally measured acoustic emissions during the histotripsy and the thermal ablation insonations are compared with the simulations conducted in subsection 3.3.

According to the *ex vivo* PCD experimental results, two important observations can be made, as follows. During the histotripsy exposure:

- More and stronger multiple harmonics appeared;
- Higher levels of broadband emissions were observed.

Despite the computational challenges in modelling the dynamics of a single spherical bubble over 100 acoustic cycles, these experimental features are consistent with the numerical results shown in figure 14.

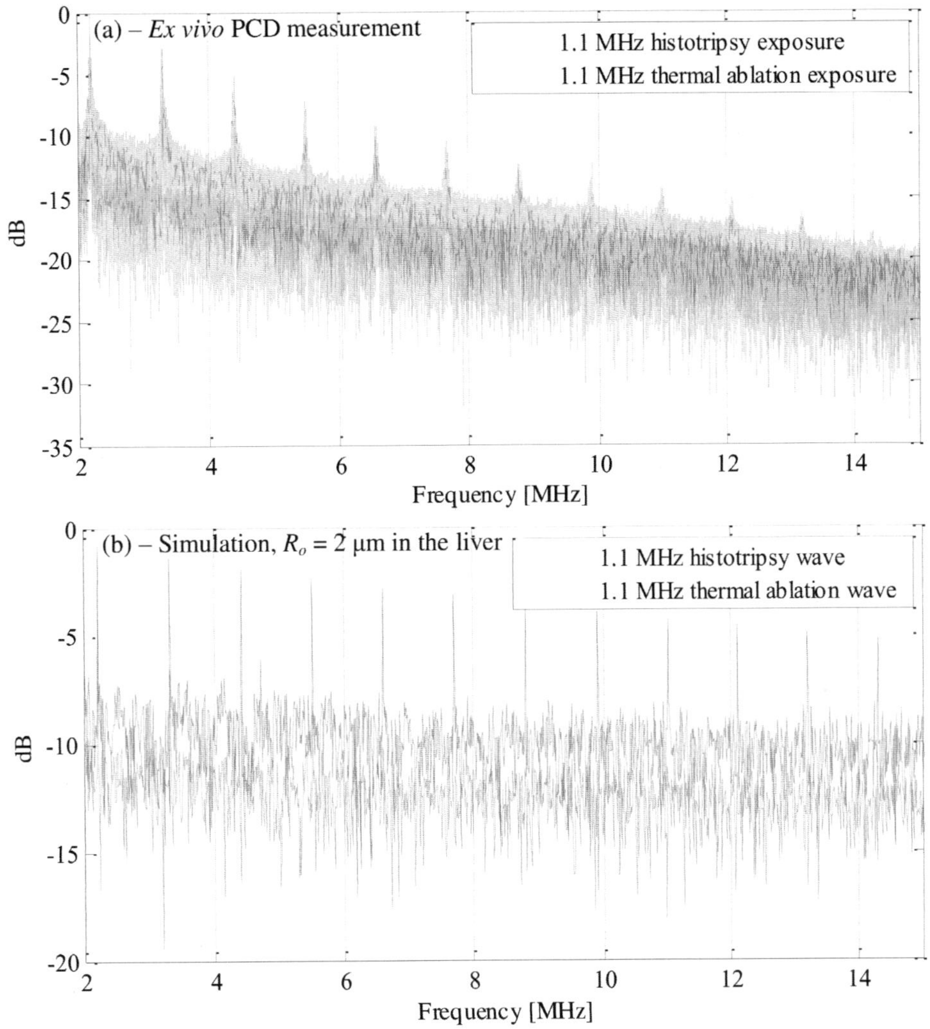

Figure 14. Comparison of the experimentally measured acoustic emissions and the numerically simulated radiated pressures in the liver. (a) is the frequency spectrum (2.0 to 15.0 MHz) obtained during both the histotripsy (blue solid line) and the thermal ablation exposure (red solid line). (b) is the simulated frequency spectrum (2.0 to 15.0 MHz) of the pressures radiated by the histotripsy (blue) and the thermal ablation waveform (red) in the liver.

5. *In vivo* experiment

In this section, an *in vivo* experiment on a rat liver is described to demonstrate the feasibility of the proposed cell therapy mediated by ultrasonic histotripsy.

5.1. Experimental method

A schematic diagram of the *in vivo* experimental setup is shown in figure 15. The experiment was conducted with a customised transducer-holder filled with degassed distilled water using a peristaltic pump (Seko, PR18). An acoustically transparent window (12 μm-thick-mylar, HiFi Industrial, PMX 980) was attached to the end of the holder (in contact with the rat). The holder coupled with a 2.0 MHz

HIFU transducer (Sonic Concepts, H106) was then linked to a manual positioner. For consistency, the HIFU focus was set to 5 mm below the surface of the liver during the experiment.

The abdomen was slightly incised and then the right lobe of the liver was taken out without dissecting the hepatic portal vein (i.e., exteriorised) after anesthetisation process. Finally, the transducer-holder was placed on the surface of the liver for the experiment. After the HIFU insonation, the exposed liver was placed back into the abdomen and the suturing process was carried out. The rat was euthanized after 72 hours and morphological and histological analyses using conventional staining techniques (hematoxylin and eosin) were conducted on the liver tissue.

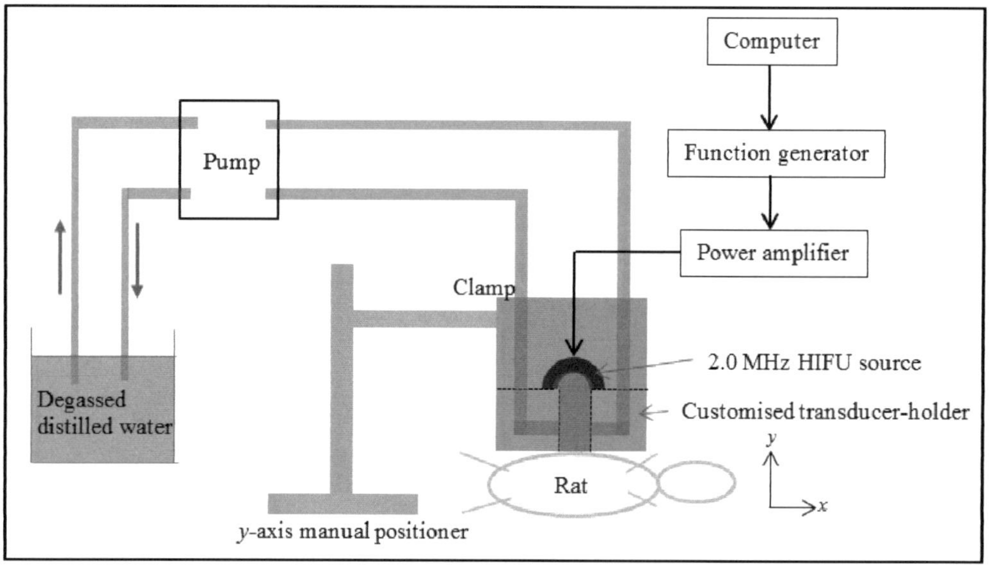

Figure 15. A schematic diagram of the *in vivo* experimental set up.

5.2. *In vivo experimental protocol-2.0 MHz HIFU transducer*

The electrical power supplied to a 2.0 MHz HIFU source was set to 200 W (P_+ = 101 MPa; P_- = -16.7 MPa). The duty cycle, pulse length, pulse repetition frequency, and number of pulses used in the experiment were shown in table 6.

Table 6. HIFU histotripsy exposure setting used during the *in vivo* experiment. 10 ms ON time for a pulse.

Duty cycle [%]	Pulse length [ms]	Pulse repetition frequency [Hz]	Number of pulses
1	10	1	50

5.3. *Experimental results*

The animal was euthanized 72 hours after the experiment for conducting the morphological and histological analyses.

5.3.1. Morphological analysis. Figure 16 shows the cross sectioned cavity (filled with clotted blood as a result of the coagulation process) produced inside the liver parenchyma. The boundary of the cavity is intact, with no evidence of coagulative necrosis which means that the cavity was induced without causing thermal damage to the cavity wall. Similarly to the *ex vivo* experimental observation (figure 8), no tadpole shaped cavity is observed (i.e., no "tail" formation).

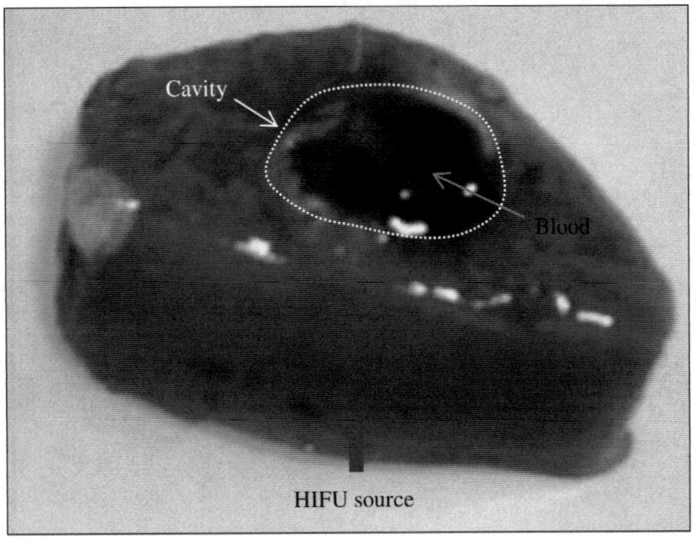

Figure 16. Cavity induced inside the liver parenchyma *in vivo*. The exposed liver was cross sectioned after 72 hours of the experiment. The cavity is filled with blood clot (due to the coagulation process) and there is no evidence of thermal damage.

5.3.2. Histological analysis. Figure 17 shows the microscopic image of the stained right lobe exposed to a 2.0 MHz HIFU field with an electrical power of 200 W (P_+ = 101 MPa and P_- = -16.7 MPa), duty cycle of 1 %, pulse length of 10 ms, pulse repetition frequency of 1 Hz and 50 pulses. There are a number of noticeable differences between the treated cavity wall and the untreated region. First, there is no sign of thermal damage around the cavity, which illustrates that the cavity was generated without causing heat damage. If thermal damage had occurred, a darker pink colour would have appeared [27]. The hepatocyte nuclei and the hepatic sinusoids are successfully located in the untreated region, whereas these are absent in the cavity as they have been destroyed by the HIFU fractionation process.

In addition, there are condensed/compressed hepatocytes (i.e., dead hepatocytes) in the cavity boundary (i.e., between the untreated region and the cavity) possibly due to the formation and expansion of the boiling bubble caused by the incident acoustic nonlinear-shocked waves.

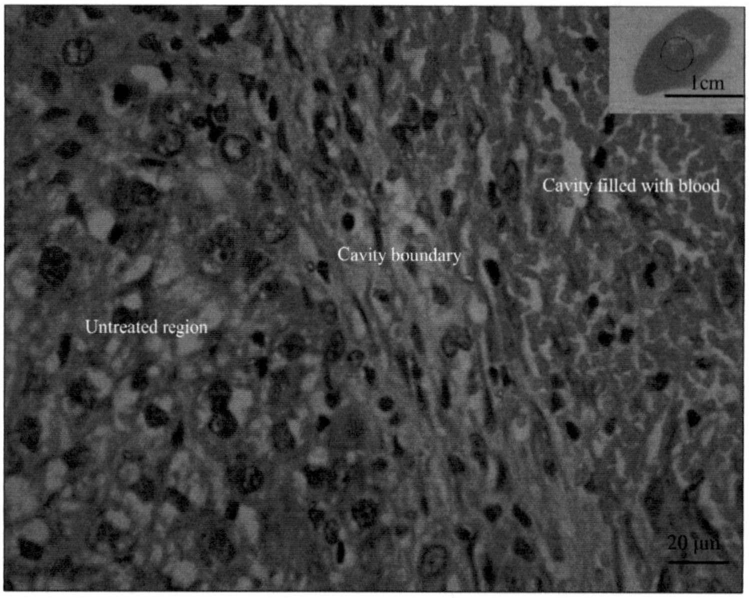

Figure 17. Microscopic image of the stained right lobe exposed to a 2.0 MHz HIFU field with an electrical power of 200 W (P_+ = 101 MPa; P_- = -16.7 MPa), duty cycle of 1 %, pulse length of 10 ms, pulse repetition frequency of 1 Hz and 50 pulses.

6. Conclusion

The numerical studies conducted with the Gilmore-Akulichev bubble model coupled with the KZK equation confirm and demonstrate that the presence of nonlinear shocked wavefront can lead to rectified bubble growth [41] (e.g., no rectified growth observed under the thermal ablation waveform which is a slightly distorted nonlinear wavefront). Furthermore, in the numerical simulation, the rectified growth can also be seen in the liver (modelled as a highly viscous human blood-like liquid); however the magnitude of the oscillations of the bubble is always small and the time to initiate the rectification process takes longer compared to that in water, because of higher viscous damping in the liver. In the future, both mass and heat transport across the bubble wall will be added into the GA model to obtain further qualitative observations.

The acoustic emissions during the thermal ablation and the boiling histotripsy exposure can be distinguished both numerically and experimentally. According to the numerical simulations, less multiple harmonics as well as lower levels of broadband emissions were observed for the thermal ablation process. These features are consistent with the PCD *ex vivo* experimental measurements.

The *in vivo* experiment performed in this work clearly illustrates the feasibility of the proposed cell therapy mediated by boiling histotripsy. As it shows that the cavity can be generated without causing coagulative necrosis, it may act as suitable housing for cell transplantation. Although hepatocyte transplantation was not conducted in this work, this study suggests that the proposed cell therapy mediated by HIFU method can potentially form an effective and safe clinical treatment for replacing and correcting disordered hepatocytes.

Hepatocytes transplantation through a cavity (i.e., ultrasound-guided needle injection) will be carried out in the near future.

References

[1] Neuberger J 2004 *Gut* **53** 759–68
[2] Yu Y, Fisher J, Lillegard J, Rodysill B, Amiot B and Nyberg S 2012 *Liver Transpl.* **18** 9–12

[3] Fox I, Roy-Chowdhury J, Kaufman S, Goertzen T, Roy-Chowdhury N and Warkentin P 1998 *N. Engl. J. Med.* **338** 935–43
[4] Horslen S, McCowan T and Goertzen T 2003 *Pediatrics* **111** 1262
[5] Russo F P and Parola M 2011 *Cytotherapy* **13** 135-44
[6] Gustafson E K, Elgue G, Hughes R D, Mitry R R, Sanchez J, Haglund U, Meurling S, Dhawan A, Korsgren O and Nilsson B 2011 *Transplantation* **91** 632
[7] Zhou H, Dong X, Kabarriti R, Chen Y, Avsar T, Wang X, Ding J, Liu L, Fox I J, Roy-Chowdhury J, Roy-Chowdhury N and Guha C 2012 *PLOS ONE* **7** 1–7
[8] Parsons J, Cain C, Abrams G and Fowlkes J 2006 *Ultrasound Med. Biol.* **32** 115–29
[9] Winterroth F, Xu Z, Wang T Y, Wilkinson J E, Fowlkes J B, Roberts W W and Cain C A 2011 *Ultrasound Med. Biol.* **37** 78–86
[10] Simon J C, Sapozhnikov O A, Khokhlova V A, Wang Y K, Crum L A and Bailey M R 2012 *Phys. Med. Biol.* **55** 8061–78
[11] Xu Z, Hall T L, Fowlkes J B and Cain C A 2007 *J. Acoust. Soc. Am.* **122** 229–36
[12] Canney M S, Khokhlova V A, Bessonova O V, Bailey M R and Crum L A 2010 *Ultrasound Med. Biol.* **36** 250–67
[13] Humphrey V F 2000 *Ultrasonics* **38** 267–72
[14] Soneson J E 2008 A user-friendly software pacakge for HIFU simulation *Proc. 8th Int. Symp. on Therapeutic Ultrasound (Minneapolis, MN)* 165–9
[15] Xu J, Bigelow T A and Nagaraju R 2013 *Trans. Ultrason. Ferroelectr. Freq. Control* **60** 1918–29
[16] Chen C, Liu Y, Maruvada S, Myers M and Khismatullin D 2012 *Phys. Med. Biol.* **57** 937–61
[17] Soneson J E and Myers M R 2007 *J. Acoust. Soc. Am.* **122** 2526–31
[18] Maruvada S, Liu Y, Pritchard W F, Herman B A and Harris G R 2011 *Phys. Med. Biol.* **57** 1–19
[19] Xu J, Bigelow T A and Riesberg G M 2012 *Ultrasound Med. Biol.* **38** 1918–29
[20] Vokurka K 1986 *Acta Acustica. Acustica* **59** 214–9
[21] Church C C 1989 *J. Acoust. Soc. Am.* **86** 215–27
[22] Ayme-Bellegarda E J 1990 *J. Acoust. Soc. Am.* **88** 1054–60
[23] Choi M J, Guntur S R, Lee K I, Paeng D G and Coleman A 2012 *Ultrasound Med. Biol.* **39** 439–48
[24] Zhu S and Zhong P 1999 *J. Acoust. Soc. Am.* **106** 3024–33
[25] Church C C, Labuda C and Nightingale K 2012 Should the mechanical index be revised for ARFI imaging? *IEEE Proc. Int. Ultrasonics Symp.* 17–20
[26] Nagoya H, Obara T and Takayama K 1995 Underwater shock wave propagation and focusing in inhomogeneous media *Proc. 19th Intl. Symp. on Shock Waves* 439–444
[27] Wang Y N, Khokhlova T, Bailey M, Hwang J H and Khokhlova V 2012 *Ultrasound Med. Biol.* **39** 424–38
[28] Khokhlova T D, Canney M S, Khokhlova V A, Sapozhnikov O A, Crum L A and Bailey M R 2011 *J. Acoust. Soc. Am.* **130** 3498–510
[29] Canney M S, Khokhlova V A, Bessonova O, Bailey M and Crum L 2009 *Ultrasound Med. Biol.* **36** 250–67
[30] Kreider W, Bailey M R, Sapozhnikov O A, Khokhlova V A and Crum L A 2011 The dynamics of histotripsy bubbles *Proc. 10th Int. Symp. on Therapeutic Ultrasound* vol 1359 427–30
[31] Lauterborn W 1976 *J. Acoust. Soc. Am* **59** 283–96
[32] Lauterborn W and Mettin R 1999 Nonlinear bubble dynamics: response curves and more. *Sonochemistry and Sonoluminescence: Proc. NATO Advanced Study Institute on Sonoluminescence and Sonoluminescence (Washington, USA, 18–29 August 1997)* eds Crum L A, Mason T J, Reisse J L and Suslick K S pp 63–72
[33] Keller J B and Miksis M 1980 *J. Acoust. Soc. Am.* **68** 628–33
[34] Yang X and Church C C 2005 *J. Acoust. Soc. Am.* **118** 3595–606
[35] Neppiras E A 1969 *J. Acoust. Soc. Am.* **46** 587–601

[36] Gae-Garreton L, Vargas-Hemandez Y, Vargas-Herrera R, Gallego-Juarez J A and Montoya-Vitini F 1997 *J. Acoust. Soc. Am.* **101** 2536–40
[37] Leighton T G 1994 *The Acoustic Bubble* (London: Academic Press)
[38] Khokhlova T D and Hwang J W 2011 *J. Gastr. Oncol.* **2** 175–84
[39] Pennes H H 1948 *Journal of Applied Physiology* **1** 93–122
[40] Tezel A and Mitragotri S 2003 *Biophysical J.* **85** 3502–12
[41] Kreider W, Maxwell A D, Khokhlova T, Simon J C, Khokhlova V A, Sapozhnikov O and Bailey M R 2013 Rectified growth of histotripsy bubbles *Proc. 21st Int. Cong. on Acoustics (Montreal, Canada, 2–7 June 2013)* vol 19 1–4

Journal of Physics: Conference Series **581** (2015) 012002 doi:10.1088/1742-6596/581/1/012002

Imaging and detection of early stage dental caries with an all-optical photoacoustic microscope

D A Hughes[1], A Sampathkumar[2], C Longbottom[3] and K J Kirk[1]

[1]. School of Physics and Engineering, University of the West of Scotland, Scotland, UK
[2]. F. L. Center for Biomedical Engineering, Riverside Research, New York, USA
[3]. Dental Innovation and Translation Centre, Kings College, London, UK

E-mail: david.hughes@uws.ac.uk

E-mail: ashwin@riversideresearch.org

Abstract. Tooth decay, at its earliest stages, manifests itself as small, white, subsurface lesions in the enamel. Current methods for detection in the dental clinic are visual and tactile investigations, and bite-wing X-ray radiographs. These techniques suffer from poor sensitivity and specificity at the earliest (and reversible) stages of the disease due to the small size (<100μm) of the lesion. A fine-resolution (600 nm) ultra-broadband (200 MHz) all-optical photoacoustic microscopy system was is used to image the early signs of tooth decay. *Ex-vivo* tooth samples exhibiting white spot lesions were scanned and were found to generate a larger (one order of magnitude) photoacoustic (PA) signal in the lesion regions compared to healthy enamel. The high contrast in the PA images potentially allows lesions to be imaged and measured at a much earlier stage than current clinical techniques allow. PA images were cross referenced with histology photographs to validate our experimental results. Our PA system provides a non-contact method for early detection of white-spot lesions with a high detection bandwidth that offers advantages over previously demonstrated ultrasound methods. The technique provides the sensing depth of an ultrasound system, but with the spatial resolution of an optical system.

1. Introduction

Dental caries remains as one of the most common oral diseases in the world [1]. It is the bacterially mediated demineralisation of the hard tissues of the teeth: enamel and dentin [2]. Under normal oral hygienic conditions, the hard tissues of the teeth (enamel and dentin) undergo a constant cycle of demineralisation and re-mineralisation. However, if the environmental pH of the tooth drops below 5.5, demineralization proceeds faster than remineralisation, interrupting the balance of this cycle. Naturally occurring bacteria in the mouth feed on the sugary remnants of food and produce acidic waste products that increase the acidity [2].

At the early stages of the disease, caries manifests as small (<100μm) sub-surface lesions, containing exogenous organic material, appearing as white 'spots' on the tooth surface (as shown in Figure 1). If undiagnosed, the lesion grows in size, spreading through the dentin layer. A cavity occurs when the lesion's surface is disrupted. After this, the process accelerates eventually producing an infection extending to the pulp chamber (which contains nerves and blood vessels) causing inflammation and pain. At this point the tooth would either be extracted, or a filling put in place.

If the caries lesion is detected at a sufficiently early stage, treatment regimes such as increased brushing and flossing or fluoride treatments, can be prescribed which can reinstate the normal mineralisation cycle and thereby halt and possibly reverse the progression of the disease. Current methods for detecting and diagnosing dental caries employing the dental explorer and X-ray radiography are subjective and not quantitative, and therefore are unreliable at the early (reversible) stages of the disease. Furthermore, the dental-explorer technique cannot be guaranteed to be non-invasive, and radiography carries safety risks regarding the use of ionizing radiation. The motivation behind this work is to develop a non-contact, non-invasive and non-ionizing technique to detect early-stage caries and to provide quantitative information for lesion evaluation.

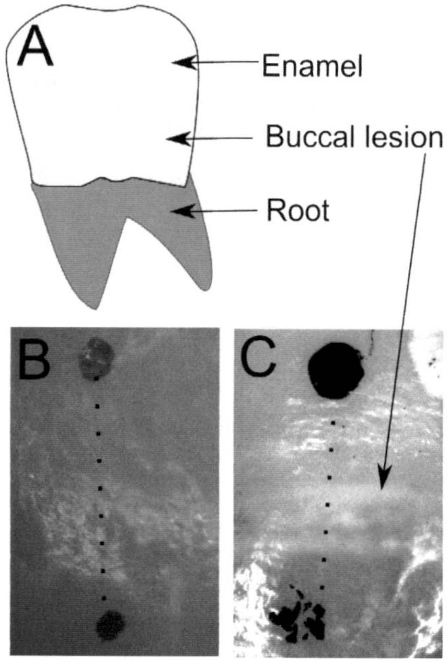

Figure 1. Diagram of a tooth showing a lesion on the buccal surface (A), and photographs of enamel surfaces showing positions of scans over (B) healthy and (C) lesioned areas. Start and end points of the scan are marked with black spots

Several ultrasonic dental imaging methods have been investigated in recent years. These range from conventional pulse-echo imaging at mid-level frequencies [3, 4, 5], CHIRP encoded imaging [6], to laser detected CHIRP encoded imaging [7]. A full review of dental uses of ultrasound is outside the scope of this article, and therefore the reader is directed towards the many comprehensive reviews of the subject [8, 9, 10]. High frequency ultrasound has recently been demonstrated for detecting dental caries [11, 10]. However, the images suffer from poor contrast between healthy and diseased enamel. In addition, conventional ultrasound also requires a coupling medium between the acoustic source and tooth which is impractical in a dental clinic.

Photoacoustic imaging (PA) is emerging as a new modality for non-invasive medical imaging and diagnosis [12, 13, 14, 15]. The PA effect is the generation of ultrasonic waves after absorption of a pulse of light in a material. PA imaging has also been demonstrated with dental tissues [16, 17, 18, 19]. Typically, these studies use a UV light source at a wavelength strongly absorbed [19] by the enamel since these lasers tend to be available to the dental clinician for more destructive cleaning techniques.

1.1. Theory of detecting tooth decay with PA

In the work presented here, a wavelength of light is chosen that has a stronger absorption coefficient in the caries lesion than the surrounding enamel. A larger PA response is expected from white spot lesions compared to the surrounding healthy enamel due to the higher absorption producing a greater change in temperature. This can be demonstrated by considering the one dimensional temperature distribution for a material exposed to a nanosecond pulse of energy [17]:

$$T(z) = \frac{\mu_a \psi(z)}{\rho C_p} \tag{1}$$

Where $T(z)$ is the temperature at depth z, μ_a is the optical absorption coefficient, $\psi(z)$ is laser energy density (the fluence) at z, ρ is the density of the material, and C_p is the specific heat capacity of the material. The laser energy density can be assumed to follow the Beer Lambert Law, such that

$$\psi(z) = E_0 e^{-\mu_a z} \tag{2}$$

Where E_0 is the energy at the surface of the material and μ_a and z as before. For completeness, the heat equation should be considered to take into account the thermal diffusion during the excitation. However, for nano-second pulses, it is found that the diffusion acts on a much longer time scale and can be neglected.

$$\frac{\partial T}{\partial t} = k_e \bigtriangledown^2 T + h(t) \tag{3}$$

Where T is the temperature, k_e is the thermal diffusivity coefficient, and $h(t)$ is a function describing the input of energy (in this case, laser pulse, $\psi(z)$ convolved with a Gaussian pulse shape). From these equations, it is possible to model the relative PA signals generated from healthy enamel and white spot lesions by incorporating values found in the literature (Shown in table 1). The results of this are shown in figure 2 where the PA signal from a white spot lesion is found to be around double that of a signal excited in healthy enamel. The data shown in Figure 2 is the derivative of the pressure, simulating the output from the interferometer used in the experimental system shown in Figure 2.2. This simplistic treatment only takes absorption into account but it is expected that the higher scattering co-efficient of white spot lesions to enamel will increase the magnitude of the Fluence term ($\phi(z)$) which will also increase the generated photoacoustic signal[17].

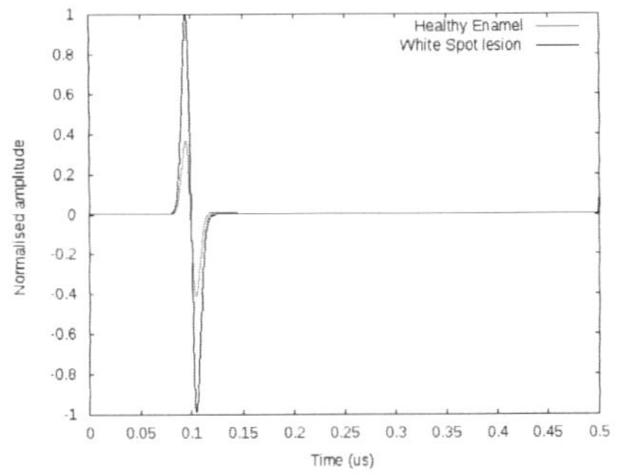

Figure 2. Analytically modelled PA signals excited in healthy enamel and white spot lesions.

Parameter	Healthy enamel	White spot lesion	Ref
μ_a (mm^{-1})	0.02	0.04^a	[20]
ρ (gcm^{-3})	2.65-2.89	1.48-2.0	[21]
C_p $(Jkg^{-1}K^{-1})$	720	945^b	[22]
k_e (m^2s^{-1})	3.22×10^{-7}	3.81×10^{-7}	[23, 24]

^{a}Estimated value, ^{b}calculated from parameters in [22].

Table 1. Optical and thermal parameters of dental tissue.

The PA approach is improved by adopting an all-optical detection technique to eliminate the need for a coupling agent, similar to the work of [18, 19] but with a much greater detection bandwidth to resolve the caries lesions for imaging purposes. This paper describes PA imaging of early-stage caries using an all optical PA microscope. PA signals are generated using a Nd:YAG (Neodymium-doped Yttrium Aluminum Garnet) laser operating at 532 nm with a 5ns pulse duration. The light-induced broadband ultrasound wave is detected at the surface of the tooth with a path-stabilised Michelson interferometer. There are several methods to produce 2D images from PA signals [25, 26, 27]. In this work, the k-wave method is applied using with the K-Wave toolbox[28] in Matlab. The reconstructed PA images of ex-vivo tooth samples containing lesions demonstrate the high sensitivity and feasibility of this technique for detecting and evaluating early stage caries.

2. Materials and Methods

2.1. Sample Preparation

Teeth extracted during orthodontic procedures in a dental clinic were stored in phosphate-buffered saline solution containing thymol to prevent bacterial growth. Samples were visually inspected under bright-field microscopy and regions presenting white spot lesions in the buccal (cheekside) region were identified for scanning under the all-optical PA microscope as shown in Figure 1A. No lapping or polishing was performed on the samples in order to demonstrate the feasibility of the PA microscope in a clinical setting.

The scan area for PA imaging of the tooth was demarcated using a marker pen as shown in Figures 1B and 1C. A linear scan was performed from top to bottom between the marked regions on the tooth. These locations were chosen so that the scan would begin in an area of healthy enamel, travel through the lesion and end in an area of healthy enamel as indicated in Figure 1C. Control samples that showed no signs of disease were imaged similarly using the PA microscope as illustrated in Figure 1B.

2.2. PA Imaging

A schematic diagram of the PA microscope is shown in Figure 2. A frequency-doubled Nd:YAG pulsed laser operating at 532 nm served as a pump source to excite PA effects in tooth. The laser produced 5-ns pulses at 10 Hz with a peak energy of 25 mJ/pulse. The pump light pulses were delivered to the tooth using a liquid-filled light guide in a backward-mode configuration. The sample was mounted on a three-axis, precision stage that allowed 2D scanning of the sample.

The PA-induced surface motion of the sample was detected by a path-stabilized Michelson interferometer [29]. The interferometer used a frequency-doubled Nd:YAG continuous-wave laser with a peak power of 300 mW operating at 532 nm as the detection beam. The output of the laser was collimated and split into a signal beam (i.e. the probe beam) and a reference beam. The probe beam was directed to a mirror and focused to a point on the tooth by the 50X objective. The spot size of the probe beam on the sample surface was measured to be 0.484 μm at the full-width half-maximum point. The probe beam was phase modulated by the motion of the sample surface and was reflected back to the beam splitter where it interfered with a reference beam that was reflected from a mirror mounted on a piezo-stack. The interfering beams were directed to a high-speed photo-detector with GHz bandwidth. The output of the photo-detector was amplified and directed to an oscilloscope that measured the time-domain signal with respect to a reference trigger derived from the pulsed pump laser.

PA data were collected along a linear scan of 3mm (Figure 1) with a step size of 10 μm. At each scan point, the tooth was moved to the focal plane of the objective to provide maximal detection sensitivity.

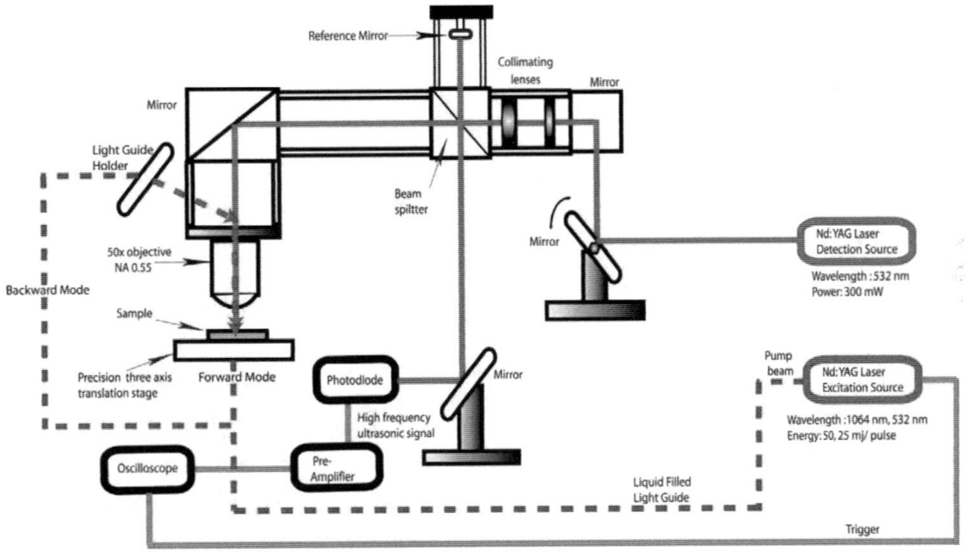

Figure 3. Diagram of all-optical PA microscope system.

2.3. Image Reconstruction

The measured time-domain signal is the superposition of the PA signals originating from all locations within the illuminated region of the tooth that contain light-absorbing chromophores. Materials with a higher optical absorption, such as a lesion, excited at a particular optical wavelength (532 nm) will generate a stronger PA signal. Spatial distributions and relative concentrations of various chromophores can be inferred by employing time-reversal processing of measured time-domain signals[26]. The reconstructed time-domain signals provide the time of arrival of the optically generated acoustic waves over the tooth surface. By knowing the speed of sound in the tooth (a value of $6500ms^{-1}$[30] is used), the acoustic signals can be spatially resolved and back-projected to form a 2D or 3D map of optical-absorption properties. Many sophisticated algorithms, like the inverse spherical-Radon transform or Fourier reconstruction[31], are available to reconstruct the distribution of optical absorbers [25].

In this study, a k-space reconstruction method [26, 32],implemented with the K-wave toolbox in Matlab [28], was used to obtain the absorption properties. This method solves the inverse problem of PA imaging by using back-propagating algorithms to reconstruct the absorbed optical energy distribution from the time-dependent PA signals. In most cases, this distribution depicts the local absorption coefficient at a particular optical wavelength concurrently with the incident light distribution and hence it provides an indirect representation of the optical properties of the tooth. In order to map the true morphology and any potential pathology, quantitative absorption information, independent of the fluence and wavelength of the light source, is required. The k-space reconstruction algorithm recovers this information by using an iteratively applied forward model of light transport[27]. This algorithm produces a spatial distribution of the absorption coefficient, which is directly related to the spatial distribution of absorbing tissue chromophores and produces quantitative absorption maps that are directly related to specific chromophores.

The PA signal generated at the laser focal point propagates in all directions through the tooth, with reflected ultrasound wavelets arriving at the surface later in time to the initial light pulse. Thus, the displacements measured by the interferometer can also result from ultrasound reflected from an unknown number of reflectors in the tooth.

3. Results

Representative ultrasonic RF-signal data collected from areas of healthy and lesioned enamel are shown in Figure 4A. These are normalised to the white spot lesion signal to emphasise the increased return signal and to allow for easy comparison with the modelled results in figure 2 where a similiar increase of signal is found. These data alone could be used to detect a lesion on the surface of the tooth. However, no quantitative data are provided on the depth to which the lesion has grown and therefore it is useful to reconstruct the PA signal.

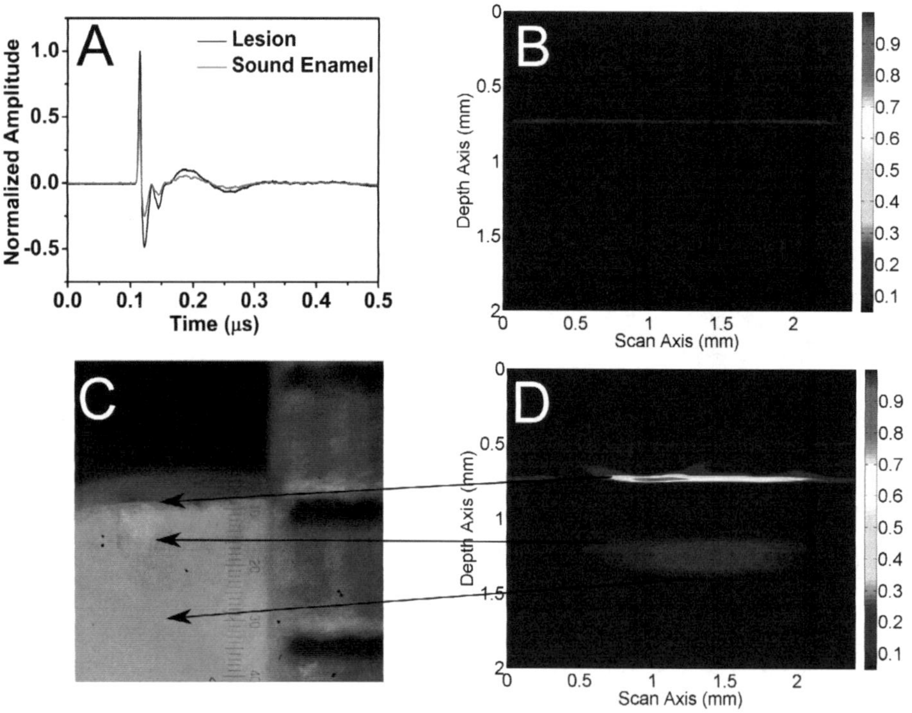

Figure 4. Raw RF-signal data collected from the PA microscope is shown in A, with the reconstructions of healthy enamel and enamel presenting a lesion shown in B and D respectively. A photograph of the hemisected lesion presenting sample is shown in C.

The time-reversal reconstructions of the PA scans are shown in Figure 4B and 4D. Both images have been normalised to the maximal amplitude, and both demonstrate the larger PA signal emerging from the lesion. Figure 4B shows an expanse of blue with a subtle PA signal (not more than 0.4 in the normalised scale), whereas Figure 4D shows a clear area of increased signal (¿0.6 in the normalised scale) coming from the white spot lesion. The increase of PA signal at the lesion is also apparent in Figure 4D where the signal drops off by an order of magnitude in the healthy enamel either side of the lesion.

We also observed some features beneath the lesion in Figure 4D that do not appear under the healthy enamel. The dimensions of this feature correspond to the geometry of the lesion shown in Figure 4C. The depth of the lesion in the PA image is found to be 0.4 mm, compared to 0.38 mm in the histological image. The thickness of the enamel (measured from the surface to the enamel dentine junction EDJ) was 0.6 mm in the PA image, and 0.84 mm in the histological image. The slight mismatch in the measured dimensions, such as the thickness of the lesion, and the distance to the enamel-dentine junction, is most likely attributed to the defined speed of sound value in the k-wave toolbox ($6500ms^{-1}$). The speed of sound in enamel has been shown to vary with depth, and the lesion itself will exhibit a lower velocity. From the measurements presented here, there is a 3% error in the measurement of the lesion thickness in the image, which corresponds closely to the measurements of changes in velocity of sound due to a white spot lesions [33]. The error in the measurement to the enamel dentine junction is much more pronounced at 30%, and while the velocity will change with depth (due to the structure of the enamel close to the EDJ)[33], further experiments and analysis will have to be carried out to rule out other factors such as the very small amplitude of the propagating ultrasound wavelet, the high attenuation in the dental hard tissues and the viewing aperture of the k-wave algorithm.

The amplitude of the PA signal was maximal at the surface of the tooth where the laser was focused. In order to reconstruct an image of the sub-surface features, the generated ultrasound wave must propagate through the tooth and be reflected from the boundaries between the different materials, e.g. enamel, lesioned enamel and dentine. The key parameter in these reflections is acoustic impedance, with healthy enamel and dentine having values of 16.5MRayls and 8MRayls[34] respectively. Early stage lesions have a slightly lower acoustic impedance than enamel, with the value corresponding to the degree of the disease. Reflected PA waves therefore have a much lower amplitude than generated PA waves, and consequently reflected waves are detected only if they have sufficient amplitude to cause an optical path-length difference in the interferometer.

For imaging, an ideal RF dataset for the time-reversal algorithm would be obtained using a tomographic configuration utilising detectors placed all around the tooth to provide a full viewing aperture. However, in the system presented here, the detection only occurs at the tooth surface and therefore only waves traveling in a specific direction are detected.

4. Conclusions

The all-optical PA microscope appears to be a promising tool for detecting early-stage caries on the surface of teeth. Lesions can be detected because of their larger PA signal amplitudes compared to the PA signal amplitudes generated by healthy enamel. Furthermore, the depth of a lesion can be measured from a time-reversal reconstruction of the RF data. The method has a number of advantages over traditional ultrasound methods such as improved contrast between lesioned and healthy enamel and no requirement for a coupling medium between the source and the tooth. These studies are now continuing with emphasis on miniaturising the system and improving our understanding of the k-wave algorithms for our application's data.

References

[1] Bagramian R A, Garcia-Godoy F and Volpe A R 2009 *American journal of dentistry* **22** 3–8 ISSN 0894-8275
[2] Selwitz R H, Ismail A I and Pitts N B 2007 *Lancet* **369** 51–9 ISSN 1474-547X URL `http://www.ncbi.nlm.nih.gov/pubmed/17208642`
[3] Ślak B, Ambroziak A, Strumban E and Maev R G 2011 *Acta of bioengineering and*

biomechanics / Wrocaw University of Technology **13** 65–70 ISSN 1509-409X URL http://www.ncbi.nlm.nih.gov/pubmed/21500765

[4] Hughes D A, Girkin J M, Poland S, Longbottom C, Button T W, Elgoyhen J, Hughes H, Meggs C and Cochran S 2009 *Ultrasonics* **49** 212–8 ISSN 1874-9968 URL http://dx.doi.org/10.1016/j.ultras.2008.08.007

[5] Culjat M, Singh R S, Yoon D C and Brown E R 2003 *IEEE transactions on medical imaging* **22** 526–9 ISSN 0278-0062 URL http://ieeexplore.ieee.org/articleDetails.jsp?arnumber=1200928

[6] Harput S, Evans T, Bubb N and Freear S 2011 *IEEE transactions on ultrasonics, ferroelectrics, and frequency control* **58** 2096–106 ISSN 1525-8955 URL http://ieeexplore.ieee.org/xpls/abs_all.jsp?arnumber=6039999' escapeXml='false'/>

[7] Santos S D, Domenjoud M and Prevorovsky Z 2010 *Physics Procedia* **3** 913–918 ISSN 18753892 URL http://dx.doi.org/10.1016/j.phpro.2010.01.117

[8] Ghorayeb S R, Bertoncini C A and Hinders M K 2008 *IEEE transactions on ultrasonics, ferroelectrics, and frequency control* **55** 1256–66 ISSN 1525-8955 URL http://www.ncbi.nlm.nih.gov/pubmed/18599413

[9] Xiang X, Sowa M G, Iacopino A M, Maev R G, Hewko M D, Man A and Liu K Z 2010 *Journal of periodontology* **81** 186–98 ISSN 1943-3670 URL http://www.ncbi.nlm.nih.gov/pubmed/20151796

[10] Marotti J, Heger S, Tinschert J, Tortamano P, Chuembou F, Radermacher K and Wolfart S 2013 *Oral surgery, oral medicine, oral pathology and oral radiology* **115** 819–832 ISSN 2212-4411 URL http://www.ncbi.nlm.nih.gov/pubmed/23706922 http://www.sciencedirect.com/science/article/pii/S2212440313001727

[11] Hughes D A, Girkin J M, Poland S, Longbottom C and Cochran S 2009 *2009 IEEE International Ultrasonics Symposium* (IEEE) pp 1–3 ISBN 978-1-4244-4389-5 URL http://ieeexplore.ieee.org/xpls/abs_all.jsp?arnumber=5441480 http://ieeexplore.ieee.org/lpdocs/epic03/wrapper.htm?arnumber=5441480

[12] Kruger R A, Liu P, Fang Y R and Appledorn C R 1995 *Medical physics* **22** 1605–9 ISSN 0094-2405 URL http://www.ncbi.nlm.nih.gov/pubmed/8551984

[13] Oraevsky A A, Jacques S L, Esenaliev R O and Tittel F K 1994 *Society of Photo-Optical Instrumentation Engineers (SPIE) Conference Series* **2134** 122 ISSN 0277-786X URL http://adsabs.harvard.edu/abs/1994SPIE.2134A.122O

[14] Zhang H F, Maslov K, Stoica G and Wang L V 2006 *Nature biotechnology* **24** 848–51 ISSN 1087-0156 URL http://www.ncbi.nlm.nih.gov/pubmed/16823374

[15] Zhang E, Laufer J and Beard P 2008 *Applied Optics* **47** 561 ISSN 0003-6935 URL http://ao.osa.org/abstract.cfm?URI=ao-47-4-561

[16] Wang H c, Fleming S, Lee Y c, Swain M and Law S 2011 *Biomedical optics express* **17** 345–355

[17] Li T and Dewhurst R J 2010 *Journal of Physics: Conference Series* **214** 012028 ISSN 1742-6596 URL http://iopscience.iop.org/1742-6596/214/1/012028

[18] Wang H c, Fleming S, Lee Y c, Law S and Swain M 2009 *Optics express* **17** 157–163

[19] Blodgett D W 2003 *The Journal of the Acoustical Society of America* **114** 542 ISSN 00014966 URL http://scitation.aip.org/content/asa/journal/jasa/114/1/10.1121/1.1578080

[20] Spitzer D and Bosch J T 1975 *Calcified tissue research* **17** 129–37 ISSN 0008-0594 URL http://www.ncbi.nlm.nih.gov/pubmed/1139366

[21] Huang T T Y, Jones A S, He L H, Darendeliler M A and Swain M V 2007 *Journal of dentistry* **35** 737–43 ISSN 0300-5712 URL http://www.sciencedirect.com/science/article/pii/S0300571207001133

[22] Hellen A 2010 *Quantitative Evaluation of Simulated Enamel Demineralization and Remineralization Using Photothermal Radiometry and Modulated* Ph.D. thesis University of Toronto

[23] Minesaki Y 1990 *Shika zairy, kikai = Journal of the Japanese Society for Dental Materials and Devices* **9** 633–46 ISSN 0286-5858 URL http://www.ncbi.nlm.nih.gov/pubmed/2134829

[24] Xu H C, Liu W Y and Wang T 1989 *Australian dental journal* **34** 530–5 ISSN 0045-0421 URL http://www.ncbi.nlm.nih.gov/pubmed/2619623

[25] Xu M and Wang L 2005 *Physical Review E* **71** 016706 ISSN 1539-3755 URL http://link.aps.org/doi/10.1103/PhysRevE.71.016706

[26] Cox B T and Beard P C 2005 *The Journal of the Acoustical Society of America* **117** 3616 ISSN 00014966 URL http://scitation.aip.org/content/asa/journal/jasa/117/6/10.1121/1.1920227

[27] Cox B T and Arridge S R 2005 *SPIE BIOS* January pp 49–55

[28] Treeby B E and Cox B T 2010 *Journal of biomedical optics* **15** 021314 ISSN 1560-2281 URL http://biomedicaloptics.spiedigitallibrary.org/article.aspx?articleid=1103324

[29] Sampathkumar A, Chitnis P V and Silverman R H 2014 *SPIE BiOS* ed Oraevsky A A and Wang L V (International Society for Optics and Photonics) p 89432Y URL http://proceedings.spiedigitallibrary.org/proceeding.aspx?articleid=1840446

[30] Huysmans M C and Thijssen J M 2000 *Journal of dentistry* **28** 187–91 ISSN 0300-5712 URL http://www.ncbi.nlm.nih.gov/pubmed/10709340
[31] Jaeger M, Schüpbach S, Gertsch A, Kitz M and Frenz M 2007 *Inverse Problems* **23** S51–S63 ISSN 0266-5611 URL http://stacks.iop.org/0266-5611/23/i=6/a=S05
[32] Cox B T, Kara S, Arridge S R and Beard P C 2007 *The Journal of the Acoustical Society of America* **121** 3453–64 ISSN 1520-8524 URL http://www.ncbi.nlm.nih.gov/pubmed/17552697
[33] Maev R G, Denisova L A, Maeva E Y and Enissov A A 2002 *Ultrasound in medicine & biology* **28** 131–136
[34] Culjat M O, Goldenberg D, Tewari P and Singh R S 2010 *Ultrasound in medicine & biology* **36** 861–73 ISSN 1879-291X URL http://www.sciencedirect.com/science/article/pii/S030156291000075X

Ultrasonic stimulation of peripheral nervous tissue: an investigation into mechanisms

C J Wright[1], J Rothwell[2] and N Saffari[1]

1. Department of Mechanical Engineering, University College London, Gower Street, London, WC1E 6BT, UK
2. UCL Institute of Neuroscience, Queen Square, London, WC1N 3BG

E-mail: ucemcjw@live.ucl.ac.uk

Abstract. Neuro-stimulation has wide ranging clinical and research potential but this is currently limited either by low resolution, penetration or by highly invasive procedures. It has been reported in previous studies that ultrasound is able to elicit a neuro-stimulatory effect at a higher resolution than other non-invasive approaches but both the underlying mechanism that makes this possible and the practical details of how it can be implemented are still poorly understood. The current study has identified the main issues that need to be resolved in the field, proposing several different approaches to tackling these areas. An isolated *in vitro* peripheral nerve bundle was chosen as a simple model to demonstrate and investigate the neuro-stimulatory effect after preliminary results showed successful stimulation in a skin-nerve preparation. Early results from the nerve bundle show successful neurostimulation, indicating that structures in the peripheral nerve axon are sensitive to ultrasound. Further research using this model should reveal more precisely what structures are being affected and how to optimise the effect, helping to inform the design of future procedures and devices used in *in vivo* applications.

1. Introduction

Several effective medical techniques already exist for stimulating the Central Nervous System (CNS). The FDA has recently approved DBS (Deep Brain Stimulation) for various mental health and neurodegenerative disorders [25]. Transcranial Magnetic Stimulation (TMS) has been approved for treatment of depression and for VNS (Vegas Nerve Stimulation) and is being assessed for epilepsy and depression [10]. Transcranial ultrasound theoretically has the potential to replace or improve all of these treatments, solving many of their drawbacks. Low Intensity, Low Frequency Ultrasound (LILFU) is well placed to make rapid changes to clinical practice with the widespread use of ultrasound (US) in hospitals, High Intensity Focused Ultrasound (HIFU) devices especially can be adapted to produce the LILFU stimulation protocols allowing potentially rapid transition from the lab to the clinic. Indeed with the medical safety record of therapeutic and imaging ultrasound protocols LILFU is not expected to be dangerous and cutting edge researchers are already testing ultrasound neuromodulation protocols in humans [18].

The use of US for neuromodulation has been investigated sporadically for over 60 years but is only recently gaining significant momentum as a viable research area. Initial studies by Fry and others showed a variety of different US effects on neural tissue including potentiation and inhibition [7, 37]. The current research field can be divided into those studying peripheral nerve stimulation and others targeting central nervous structures, with the focus in both being on *in vivo* work.

The first study to re-ignite interest in ultrasound neurostimulation of the CNS was in 2008 [34], shortly followed by another *in vitro* study in 2009 [24]. This sparked a series of *in vivo* studies demonstrating the efficacy of ultrasound neurostimulation by targeting the motor cortex of mice and rabbits [32, 39, 20, 14, 40, 23].

By contrast the studies on the peripheral nervous system only show much success when targeting specifically mechanosensitive structures such as the skin [18, 8, 24, 17, 4] or neuropathic tissue [33]. Attempts at inducing Action Potentials (APs) by specifically targeting nerve axon have been hitherto mostly unsuccessful, instead resulting in inhibition only [19, 11], although modulation of APs stimulated electrically is easily observed [29, 22, 31]. Only one study has reported successful US stimulation of a peripheral nerve trunk [13] though their stimulus was transmitted through brain tissue, which given reflections that inevitably occur within a rat's skull, raises questions about their stimulation specificity.

In almost all of the literature, in both peripheral and central nervous studies, the theme follows that very little to no effort has been made to control or correct for reflections of ultrasound. This leads to mistrust of stated values of intensity and pressure and when combined with the variety of models and often conflicting results, studies become very difficult to compare. The mixed findings in the literature have even on occasion called into question existence of the stimulatory effect itself [2, 8]. The purpose of this research is to gain a better understanding of how ultrasound affects nervous tissue and create a hypothesis, backed by experimental evidence for how the mechanical forces of ultrasound are transduced into propagating electrical activity in the nerve fibres.

1.1. Biological Target Structure

Neurons are cells that have highly electrically polarised membranes; at rest their interior is highly negative with respect to the positive outside and they fire by depolarising the membrane in a wave that travels along the nerve. This electrical gradient is maintained by the cell by pumping charged ions in antiport against their concentration gradient. The major ions involved along the nerve axon are sodium and potassium. Studies reporting US neurostimulation by targeting the motor cortex of mammals [39, 34] have no way of knowing which part of the neuron is responding to ultrasound if indeed the effect is region specific. The wide variety of structures in different regions of a neuron makes it difficult to predict exactly how the ultrasound energy is being transduced into a cellular response. However, as one of the most essential parts of a nerve depolarisation response, ion channels themselves are a likely candidate.

Opening or closing of ion channels will alter their radius or hydrophobic thickness through conformational change, meaning that all of them will be to some extent sensitive to mechanical forces through the surrounding membrane or other sources, known as mechanosensitivity [35]. The simplest theory for US neurostimulation therefore is that ion channels are both the direct receiver of the US force and the only biological effectors involved in the AP generation.

Ion channels shown to exhibit particular sensitivity to mechanical stimulation include non-selective cation superfamily, two pore domain potassium channel superfamily, amiloride sensitive sodium channel superfamily and voltage gated ion channels [12]. The proposed leg nerve model used in this study will contain two pore potassium leak channels that help maintain resting potential and are constitutively active. It will also contain the voltage gated sodium and potassium channels which depolarise the membrane when activated. It is hoped that future work using the current model will be able to determine the extent to which each of these channel types are affected by US.

1.2. Energy Transduction Mechanism

The 5 influences of US propagation through living tissue are outlined by Wulff et al. in 1951 [38] as: "(1) heating caused by absorption of acoustic energy; (2) periodic pressure changes; (3) radiation pressure; (4) streaming or flow in viscous media; and, (5) high temperatures and pressures associated with cavitation, defined as the formation of holes in liquid media". Numbers 1 and 5 are usually discounted in recent studies due to the low average intensities making the predicted heating negligible and the chance of inertial cavitation low. It is possible however that stable cavitation (oscillation and growth/shrinkage without collapse) of pre-existing bubbles may be occurring.

Despite the passage of more than 60 years it has still not been determined which of the remaining mechanisms are the primary source of bio-effects in nerves. Of course all of them will be occurring to some extent in the complex environment of tissues but the importance of each will change based upon the parameters of the US stimulus being applied and the structure of the tissue. Hence the community must focus on a tight set of parameters which are optimum for eliciting certain desired forces so that its effect in different tissues can be determined. In the case of neurostimulation no optimum or desirable characteristics of the US stimulus have been arrived at by any consensus in the past literature which leaves the conjecture about the mode of action and method of energy transduction very open.

1.3. In Vitro Nerve Model

In the current study the effect of US on the excitability of peripheral crab leg nerves is tested. This model was chosen for its simple, well characterised nerve bundle [9] that is easy to extract and take recordings from. The relative biological simplicity of this nerve bundle together with a well characterised US field that is not distorted by obstructions or overlaying tissue is intended to give the current study clear and repeatable results where independent variables can be isolated and definite conclusions reached with confidence. High accuracy of US field characterisation is made possible by the lack of any solid tissue depth, allowing accurate modelling of forces on the micro and macro scale which can be compared between different studies and biological models.

Along with the benefits of an *in vitro* model, the nerve bundle will be exposed to stresses and temperature changes not expected *in vivo* which may affect sensitivity to US bio-effects. Also, the differences between central and peripheral nerves such as channel subtypes, supporting tissues and morphology mean that this model will not be used in direct comparison to recent CNS stimulation studies. However, as the model only involves axonal tissue, the Hodgkin and Huxley model AP, axon morphology and broad ion channel types are all well conserved and consistent with unmyelinated axons in humans [29, 26].

2. Experimental set up

In order to target an *ex-vivo* nerve with a controlled and known ultrasonic intensity the organ bath was floated on a 20cm deep water bath with a 2cm thick, high density foam US absorbing layer at the bottom of the tank to reduce the effect of reflections on focal amplitude, measured to be <5% of the peak focal amplitude. The water was kept cold (0-4 °C) using ice to slow the nerve AP and to maintain the viability of the nerve bundle for longer. The ice was kept from the US propagation path by 2 sheets of stainless steel mesh (figure 1 and 2a & b).

The organ bath had to be designed and built to allow for simultaneous use of various modes of stimulation. A slot was cut into the acrylic wider than the beam profile of the PA409 transducer at the focal point (figure 3a) to allow ultrasound to pass through, preventing reflection or absorption by the chamber that might amplify or alter the effects on the nerve.

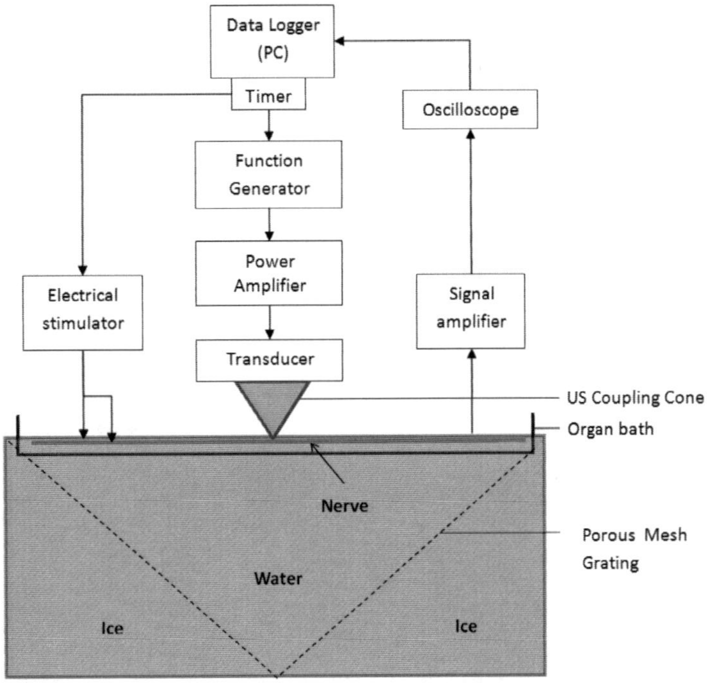

Figure 1. Schematic diagram of the experimental setup used to generate the alternate electrical and US stimului recordings seen in figure 5 and 6. Equipment details: **function generator** - Agilent 33220A. **Power amplifier** - E&I 1020L 200W class AB linear amplifier with 55dBm gain. **Transducer** - Precision Acoustics 409 PZT curved ceramic 0.67MHz HIFU. **Signal Amplifier** - WPI DAM50 differential amplifier at 100X DC gain. **Electrical stimulator** - Digitimer DS3 constant current isolated stimulator. **Oscilloscope** - Lecroy HDO6054 500MHz. The water bath and US coupling cone were filled with degassed and deionized water. The organ bath was filled with white mineral oil.

Thin mylar film (12μm) was used to cover the slot as it has an acoustic impedance closely matched to water, providing a trench to hold the nerve bundle and isolate it from the water bath. Silver foil (25μm thick) strips were placed at intervals over the mylar film and fixed into the acrylic to be used as stimulation and recording electrodes. These electrodes were chosen to be orders of magnitude thinner than the wavelength of ultrasound in water (2.3mm at 0.67MHz) to avoid unwanted absorption.Silver was necessary for a low electrical resistance through the requisite thin foil and so the metal could be silver-chloride coated to provide high resistance to current flow through the saline around the nerve.

To further increase electrical resistance between the different electrodes white mineral oil was used to fill the organ bath, acting as US transmission medium that was electrically insulating, only a thin sheath of saline was maintained around the nerve to keep it alive for the duration of the experiment. A second chilled bath filled with saline was kept next to the main test tank where the nerve was kept if not being stimulated for prolonged periods. The organ bath and all implements were cleaned and sterilised with ethanol before every use.

A water-containing coupling cone was built to provide an US propagation path from the transducer to the oil in the organ bath. The aperture of this cone was covered with mylar film and placed over the nerve between electrode 4 and 5 using a spacer attachment to ensure accurate alignment between the nerve bundle and the focal region of the US. Electrical stimulation was applied via electrode 3 using a Digitimer Ltd. DS3 model constant current isolated stimulator delivering a 0.2ms 2-20mA electrical stimulus pulse (figure 2a) at the start of the procedure and between each US stimulus. This was done to keep track of the health of the saturated Compound Action Potentials (CAPs) and therefore the state of the nerve bundle. Each stimulus event (of either modality) was separated by 3 seconds.

Ultrasound parameters were chosen based on precedent in the literature for successful neurostimulation protocols. A higher power was used here than in previously published studies due to predictions made from intensities used in peripheral nerve studies [18, 22, 7]. US parameters used are described in figure 5. Orders of parameters being tested were randomised to prevent systematic error from nerve inhibition or other effects.

To check the metrology of the pressure estimates at the focal point the US field was modelled

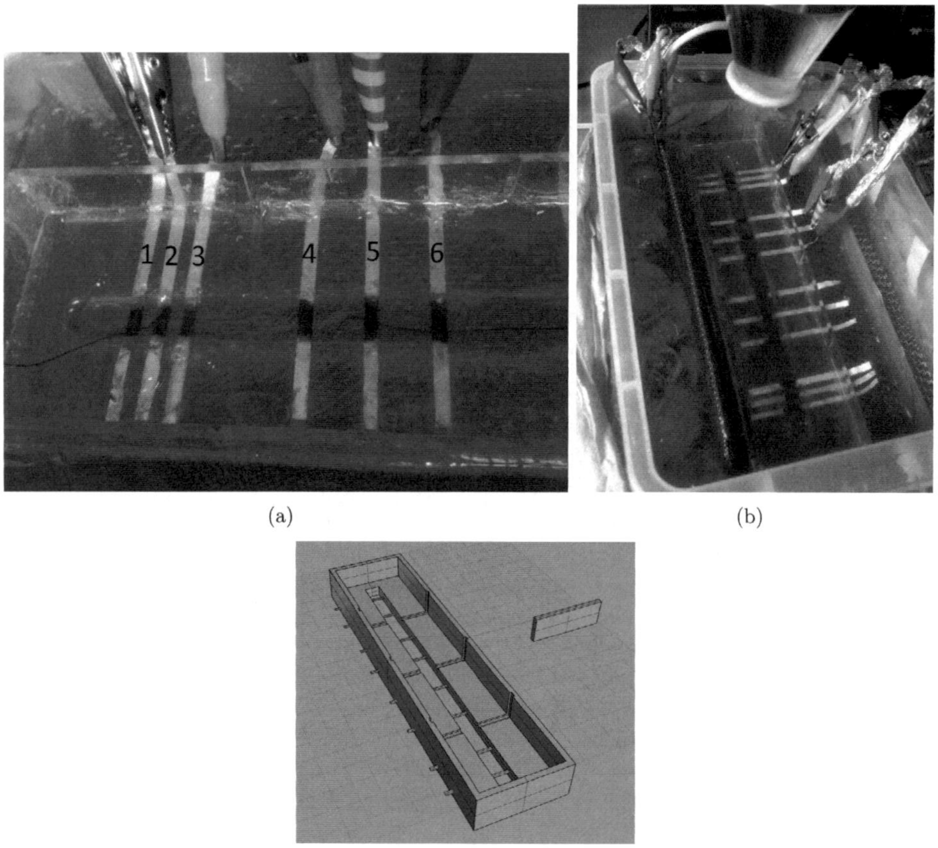

(a) (b) (c)

Figure 2. Depictions of the organ bath used to hold and stimulate the nerve via multiple modalities simultaneously. **a)** Photo of the organ bath with a nerve in place, marked by the red thread tied to each end of the nerve. Attached to the numbered electrodes are: 1) Stimulation reference 2) Stimulation active 3) Earth 4) Recording A 5) Recording B 6) Earth. **b)** Photo of the organ bath in the test tank with the ultrasound cone held above the setup. **c)** CAD diagram of the organ bath design made using Rhino $3D_{tm}$. Removable partitioning walls were built into the design to create isolated compartments when needed.

using a modified Khokhlov-Zabolotskaya-Kuznetzov (KZK) equation:

$$\partial^2_{T,z} - \frac{1}{2}\partial^2_T U^2 \beta \partial^3_T U - \gamma \Delta_y U = 0$$

where β and γ are positive constants, T is time, y and z are two linear dimension variables and U is velocity [41, 16]. This method was used to predict any non-linear propagation that may be occurring (figure 3) by modelling 50 harmonics around the fundamental frequency. At the maximum driving power used in the current study (20W) the peak positive pressure was indeed found to be higher than the peak negative (figure 3a) by a small amount along with minor harmonics generated (figure 3b). Both these effects are not seen at lower amplitude driving powers (figure 3c&d). These non-linearities should be considered in any estimation of biological forces, especially when considering medical applications with higher tissue depths. The predicted pressure field profile was found to be in good agreement with previously predicted values and hydrophone measurements (figure 3e).

After propagation through water to the focal point a single pulse was measured using a fibre optic hydrophone to demonstrate the amplitude profile that the nerve bundle would encounter (figure 4). The ramping up and trailing off of the pulse is due to the transducer's damping properties.

3. Results

Delivering a rapid pulse train of focused ultrasound directly into axonal tissue was able to produce significant excitation (figure 5) that only occurred in the first of 5 repeat stimuli. This data was collected over 5 seperate crabs (all large male *Cancer pagurus*), the large CAP response was seen in 6 out of 9 (67%) nerve extractions that responded to electrical stimulation. Of these responsive nerves only 23% of US stimulus pulse trains resulted in observable CAPs. Significant CAP events were primarily seen in response to the first US stimulus, the application of subsequent pulse trains failing to result in recorded CAP events despite the electrical stimulation showing the nerve to be functional. The electric stimulation did reveal a successive reduction in CAP amplitude between each dose of US, especially visible in figure 5a where the response to the US was also highest, although this inhibition during the 5 stimulus protocol was not always observed(figure 5c and 6f). The lower or upper thresholds for producing the neurostimulatory effect were not found in these preliminary experiments.

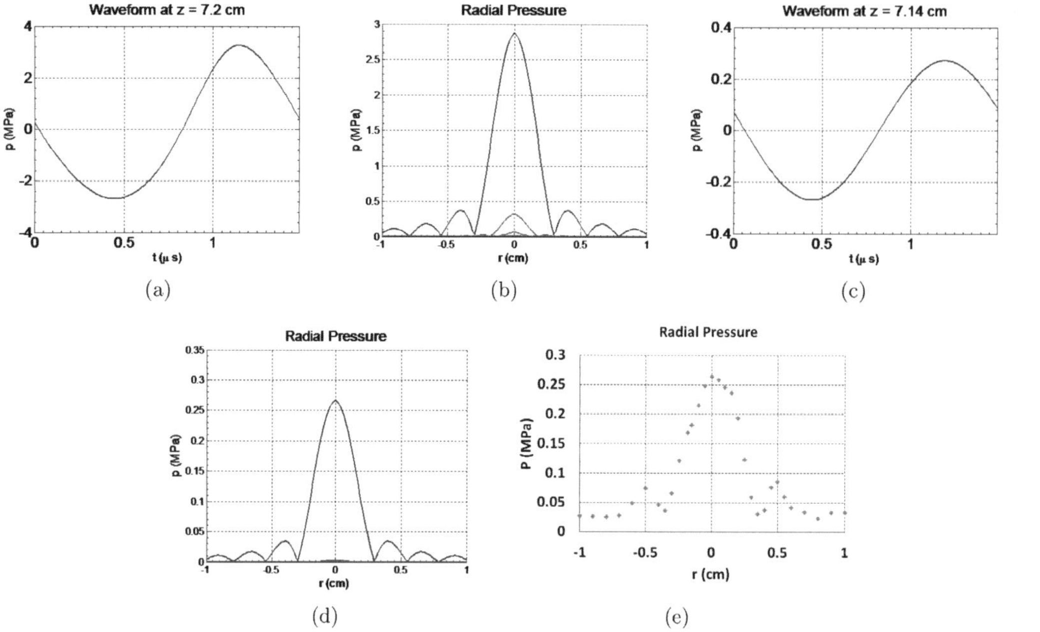

Figure 3. Objects **a** to **d** are predicted plots of the US field using the KZK acoustic model at the top and bottom end of input powers experimentally used (**a&b** - 20 Watt, **c&d** 0.17 Watt) **a&c**) Pressure oscillation at the focal point of the PA409 transducer. **b&d**) Transverse scan of the pressure amplitude profile at the modelled focal point. Second and third harmonics are displayed in green and red respectively. **e**) Pressure amplitude profile measured from the PA409 HIFU transducer using a Precision Acoustics plane tipped fibre-optic hydrophone in degassed and de-ionised water at the geometric focal point.

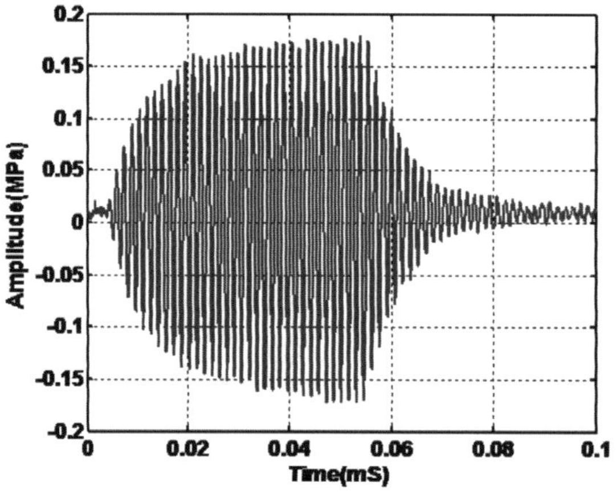

Figure 4. Acoustic pressure amplitude trace of a single pulse in the stimulation protocol. Recorded using a plane tipped fibre optic hydrophone placed in the focal region of a 0.67MHz PA407 HIFU transducer in degassed and deionised water.

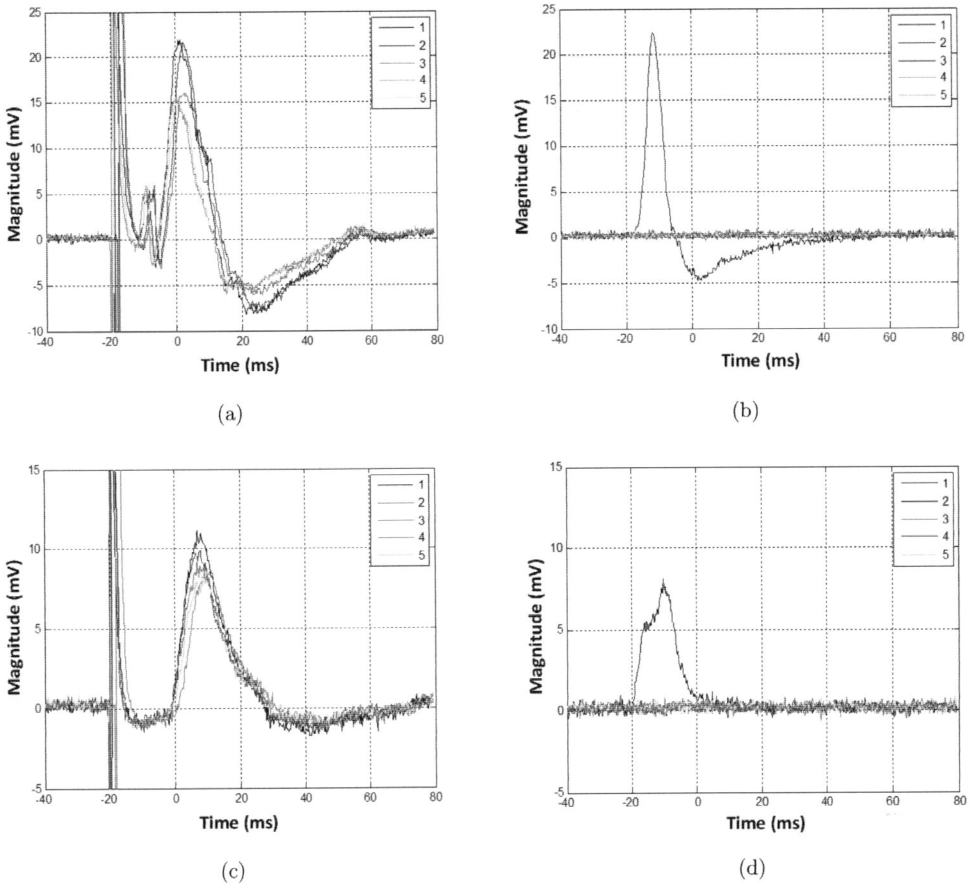

Figure 5. Recorded voltage amplitude profiles involving CAP activity from a crab leg nerve bundle using alternating electrical (200μs, 2mA) and ultrasonic stimulation (10kHz PRF, 50% duty cycle, 0.67MHz sinusoidal wave, **b)** 1MPa **d)** 1.3MPa peak focal pressure (66 W/cm^2 and 118.1 W/cm^2 I_{SPTP} respectively), total stimulus duration of 8ms (80 pulses)) of a crab leg nerve bundle with 2 seconds between each recording. Figures **a** and **c** are overlaid CAP recordings where the nerve was stimulated electrically, **a** and **c** correspond to US stimulation recordings in **b** and **d** respectively.

Recovery times of up to 20 minutes after observation of a successful US induced CAP were investigated (1,5,10 and 20 minute recovery periods) where no electrical or US stimulus was given and the nerve was not moved, the stimulus amplitude and pulse parameters were kept the same (figure 6). Recovery time was found to have no effect on susceptibility to US as no US induced CAPs were observed in 3 trials (12 total US stimuli events).

Movement of the nerve by 5mm so that a new section was under the US focus for each stimulus session was able to improve the US response reliability from 23% to 71% (total of 14 exposures at 260 W/cm^2 I_{SPTP}). This success did not appear to depend on which direction the nerve was moved.

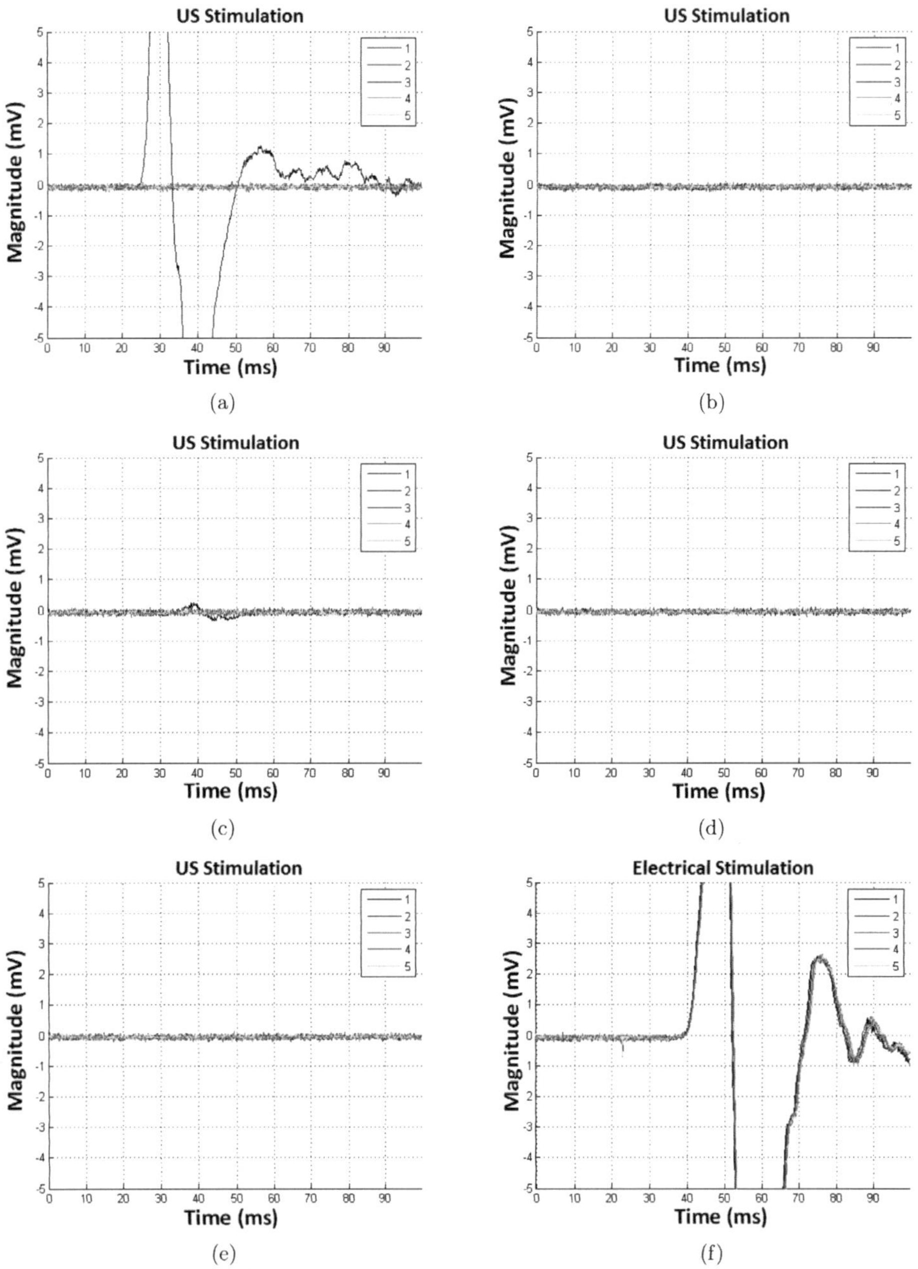

(a) (b) (c) (d) (e) (f)

Figure 6. Recorded voltage amplitude profiles from a crab leg nerve bundle using 2MPa (260W/cm^2 I_{SPTP}) peak focal pressure, otherwise identical protocol to those in figure 5. Objects **a** to **e** are nerve responses from US stimuli with longer recovery times left between each stimulus event (0,1,5,10 and 20 minutes respectively). **f** represents an example electrical stimulus profile taken alternately with the recordings in **a**. Electrical stimulus recordings were also made during US stimuli **b** to **e** but are not displayed here

4. Discussion

The electrical responses to ultrasound stimulation reported here represent a significant new finding and demonstrate the usefulness of the *ex-vivo* peripheral nerve model as a vehicle for studying the nature of the US-nerve interaction. The data found in our experiments contradict previous studies reporting that *ex-vivo* peripheral nerves cannot be stimulated by US, only modulated or inhibited [19, 22, 31, 11] and provides several novel insights into the mechanism of action, proving that purely axonal tissue and its constituents are sufficient for neurostimulation and that the reliability of the effect is high enough to simultaneously stimulate many axons in the nerve bundle. The next step using the current model is to test several of the hypotheses proposed by different groups working in the field such as ultrasound induced intra-membrane bubble cavitation [15, 28, 27] or radiation force inducing membrane tension [8, 21]. Further research using this model should reveal more precisely what structures are being affected and how to optimise the effect, helping to inform the design of future procedures and devices.

To ensure temperature was a negligible factor HIFU simulation modelling was used to determine the temperature rise at the focal maximum at the highest intensities used in this study. The predicted rise was less than 0.1°C.

In figure 5a and c the rate of degradation of electrically induced CAPs is faster than a time series control experiment with no US stimulation predicted (30% and 26% reduction over 50 seconds vs a maximum of 8% over the same time period using the control time course) especially visible in figure 5a between the 2nd and 3rd stimulus. This warrants further investigation through microscopic or histological examination of the nerve tissue after exposure to US stimulus which will be carried out in any future experiments. If no signs of damage are found then this effect is likely due to a long lasting inhibitory effect of US on nerve conduction past the point of sonication.

A location-specific self-inhibitory effect of the US mechanism appears to be at play here given the results from figure 6 where very little inhibition of the electrical stimulation was seen (36% reduced maximum CAP amplitude over 36 minutes) despite total inhibition of the US response after the first stimulation. This implies that the lack of further US response is not due to mechanical damage, as this would be expected to similarly block or slow action potentials from electric stimulation. Without mechanical damage as a source of this inhibition the authors propose that the mechanism of the observed stimulation in this experiment involves bubbles or bubble nuclei which are somehow moved or depleted during a single pulse train stimulus.

This theory is supported further by the improvement of response reliability as new sections of the nerve are targeted and by well established bubble nucleation theory that predicts US parameters very similar to the ones here and in previous studies are ideal for generating cavitational activity [6].

Although the nature and location of cavitation nuclei in tissue is debated [1], their existence has long been known [3, 5, 36], meaning the same stable cavitation stimulation mechanism predicted here may be acting *in vivo*. However, the nature of an *in vitro* model makes the presence of micro-scale bubbles more likely than *in vivo*. The presence of micro-bubbles will be quantified at a later stage and if they are found, their behaviour and forces exerted on surrounding tissues can be modelled and reproduced in their absence. Alternately their existence can be induced *in vivo* by low power continuous wave ultrasound [30].

An important attribute of the crab leg nerve model is that all the fibres in the bundle are unmyelinated which may make them more susceptible to US forces than equivalent myelinated nerve fibres that have been tested previously in mammals and amphibians [19, 22, 31, 2]. They share this attribute with the majority CNS neurones (grey matter) where the stimulatory effect has been shown to be most reliable though further study is needed to test whether the myelination itself is a significant factor.

5. Conclusions

Reported here is a novel approach to studying US neurostimulation, expressly designed to determine the nature of the US tissue interaction, not just its effects. Already the methods described above have demonstrated the US sensitivity of the *ex-vivo* peripheral nerve model, confirming that US neurostimulation mechanisms are not restricted to the structures of the CNS. Furthermore, structures within purely axonal tissue are sufficient to receive US stimulation and produce action potentials as a direct result, vastly reducing the options for the critical biological US force receivers.

Using the equipment and experiments described above it is predicted that future work will quickly be able to determine the key parameters of US that are responsible for eliciting neuromodulation. This information would have immediate value in optimising stimulation parameters used in research, beginning to provide some theoretical consensus in the field and potentially start to unify the diverse range of observed US bioeffects. Answering the two questions at the core of this study: what is the key US force mechanism? and What is the key biological structure being affected? will certainly shed much needed light on this poorly understood phenomenon.

References

[1] R. Arieli and A. Marmur. Decompression sickness bubbles: are gas micronuclei formed on a flat hydrophobic surface? *Respir Physiol Neurobiol*, 177:1923, 2011.

[2] V. Colucci. Focused ultrasound effects on nerve action potential in vitro. *Ultrasound in Medicine & Biology*, 35(10):pp.1737–1747, 2009.

[3] V.S.J. Craig. Formation of micronuclei responsible for decompression sickness. *Journal of Colloid and Interface Science*, 183:260268, 1996.

[4] D. Dalecki, S. Z. Child, C. H. Raeman, and E. L. Carstensen. Tactile perception of ultrasound. *The Journal of the Acoustical Society of America*, 97(5 Pt 1):3165–3170, May 1995.

[5] A. Evans and D. N. Walder. Significance of gas micronuclei in the aetiology of decompression sickness. *Nature*, 222(5190):251–252, April 1969.

[6] H.G. Flynn and C.C. Church. A mechanism for the generation of cavitation maxima by pulsed ultrasound. *The Journal of the Acoustical Society of America*, 76(2):505–512, 1984.

[7] W.J. Fry, V.J. Wulff, D. Tucker, and F.J. Fry. Physical factors involved in ultrasonically induced changes in living systems:i. identification of non-temperature effects. *Acoust Sot Am*, 22:867876, 1950.

[8] L.R. Gavrilov and E.M. Tsirulnikov. Focused ultrasound as a tool to input sensory information to humans. *Acoustical Physics*, 177:121, 58.

[9] G.A. Horridge and R.A. Chapman. Sheaths of the motor axons of the crab carcinus. *Quarterly Journal of Microscopical Science*, 3-105:175181, 1964.

[10] PG Janicak, Z Nahas, SH Lisanby, HB Solvason, and Sampson SM. Durability of clinical benefit with transcranial magnetic stimulation (tms) in the treatment of pharmacoresistant major depression: assessment of relapse during a 6-month, multisite, open-label study. *Brain Stimul.*, 2010.

[11] E.J. Juan, R. Gonzlez, G. Albors, M.P. Ward, and P Irazoqui. Vagus nerve modulation using focused pulsed ultrasound: Potential applications and preliminary observations in a rat. *International Journal of Imaging Systems and Technology*, 24(1):67–71, 2014.

[12] A. Kamkin and I. Kiseleva. *Mechanosensitivity of the Nervous System*. 2009.

[13] H. Kim, S.J. Taghados, K. Fischer, L. Maeng, S Park, and S.S. Yoo. Noninvasive transcranial stimulation of rat abducens nerve by focused ultrasound. *Ultrasound in Medicine & Biology*, 38(9):1568 – 1575, 2012.

[14] R.L. King, J.R. Brown, W.T. Newsome, and K.B. Pauly. Effective parameters for ultrasound-induced in vivo neurostimulation. *Ultrasound in Medicine & Biology*, 39:312331, 2013.

[15] B. Krasovitski, V. Frenkel, S. Shoham, and E. Kimmel. Intramembrane cavitation as a unifying mechanism for ultrasound-induced bioeffects. *PNAS*, 2011.

[16] V.P. Kuznetsov. Equations of nonlinear acoustics. *Soviet Physical Acoustics*, 1971.

[17] W. Lee, H. Kim, S. Lee, S. Yoo, and Y.A. Chung. Creation of various skin sensations using pulsed focused ultrasound: Evidence for functional neuromodulation. *International Journal of Imaging Systems and Technology*, 24(2):167–174, June 2014.

[18] W. Legon, A. Rowlands, A. Opitz, T.F. Sato, and W.J. Tyler. Pulsed ultrasound differentially stimulates somatosensory circuits in humans as indicated by eeg and fmri. *PLoS ONE*, 7:e51177., 2012.

[19] P.P. Lele. Effects of focused ultrasonic radiation on peripheral nerve, with observations on local heating. *Experimental Neurology*, 8(1):47 – 83, 1963.

[20] E. Mehic, J.M. Xu, C.J. Caler, N.K. Coulson, C.T. Moritz, and P.D. Mourad. Increased anatomical specificity of neuromodulation via modulated focused ultrasound. *Plos One*, 9(2):e86939, February 2014. WOS:000330631800016.

[21] M.D. Menz, . Oralkan, P.T. Khuri-Yakub, and S.A. Baccus. Precise neural stimulation in the retina using focused ultrasound. *J. Neurosci.*, 33:45504560, 2013.

[22] R.T. Mihran, F.S. Barnes, and H. Wachtel. Temporally-specific modification of myelinated axon excitability in vitro following a single ultrasound pulse. *Ultrasound in Medicine & Biology*, 16:297309, 1990.

[23] B. Min et al. Focused ultrasound modulates the level of cortical neurotransmitters: Potential as a new functional brain mapping technique. *International Journal of Imaging Systems and Technology*, 21(2):232–240, 2011.

[24] R. Muratore, J. LaManna, E. Szulman, M.S.A. Kalisz, M.S.M. Lamprecht, M. abd Simon, M.S.Z. Yu, N. Xu, B. Morrison, and E.S. Ebbini. Bioeffective ultrasound at very low doses: Reversible manipulation of neuronal cell morphology and function in vitro. *Respir Physiol Neurobiol*, page 2529., 2009.

[25] CW Olanow, MB Stern, and Sethi K. The scientific and clinical basis for the treatment of parkinson disease. *Neurology*, 2009.

[26] F.C. Page. A comparative study of five fresh-water and marine species of thecamoebidae. *Transactions of the American Microscopical Society*, 90:157173, 1971.

[27] M. Plaksin, S. Shoham, and E. Kimmel. Intramembrane cavitation as a predictive bio-piezoelectric mechanism for ultrasonic brain stimulation. *Phys. Rev. X*, 4:011004, Jan 2014.

[28] S.M. Rappaport, A.M. Berezhkovskii, J. Zimmerberg, and S.M. Bezrukov. Thermodynamics of interleaflet cavitation in lipid bilayer membranes. *Phys. Rev.*, E:87, 2013.

[29] A. Sheltawy and R.M. Dawson. The polyphosphoinositides and other lipids of peripheral nerves. *Biochem. J*, 100:1218, 1966.

[30] G.R. Ter Harr, S. Daniels, and K. Morton. Evidence for acoustic cavitation in vivo: Thresholds for bubble formation with 0.75-MHz continuous wave and pulsed beams. *IEEE Transactions on Ultrasonics, Ferroelectrics, and Frequency Control*, 33(2):162–164, March 1986.

[31] P. Tsui, S. Wang, and C. Huang. In vitro effects of ultrasound with different energies on the conduction properties of neural tissue. *Ultrasonics*, 43(7):560 – 565, 2005.

[32] Y. Tufail et al. Transcranial pulsed ultrasound stimulates intact brain circuits. *Neuron*, 66:681–694, 2010.

[33] R.E. Tych, M. Gofeld, J.G. Jarvik, M. Kliot, J.D. Loeser, A.M. McClintic, R.J. Ollos, K.D. Pederson, R.E. Sparks, G.W. Terman, and P.D. Mourad. Neuropathic tissue responds preferentially to stimulation by intense focused ultrasound. *Ultrasound Med Biol*, 39:111116, 2013.

[34] W.J. Tyler et al. Remote excitation of neuronal circuits using low-intensity, low-frequency ultrasound. *PloS One*, 3(10):3511, 2008.

[35] T. Ursell, R. Phillips, J. Kondev, D. Reeves, and P.A. Wiggins. The role of lipid bilayer mechanics in mechanosensation. *Respir Physiol Neurobiol*, 2007.

[36] R. D. Vann, J. Grimstad, and C. H. Nielsen. Evidence for gas nuclei in decompressed rats. *Undersea Biomedical Research*, 7(2):107–112, June 1980.

[37] P.D. Wall, D. Tucker, F.J. Fry, and J. Mosberg. The use of high intensity ultrasound in experimental neurology. *The Journal of the Acoustical Society of America*, 25:281285, 1953.

[38] V.J. Wulff, W.J. Fry, D. Tucker, F.J. Fry, and C. Melton. Effects of ultrasonic vibrations on nerve tissues. *Exp Biol Med*, 76:361366, 1951.

[39] S.S. Yoo et al. Focused ultrasound modulates region-specific brain activity. *NeuroImage*, 56(3):1267–1275, 2011.

[40] Y. Younan, T. Deffieux, B. Larrat, M. Fink, M. Tanter, and J.-F. Aubry. Influence of the pressure field distribution in transcranial ultrasonic neurostimulation. *Medical Physics*, 40:082902, 2013.

[41] E.A. Zabolotskaya and R.V. Khokhlov. Quasi-plane waves in the nonlinear acoustics of confined beams. *Soviet Physical Acoustics*, 1969.

Multi-resolution analysis of passive cavitation detector signals

S R Haqshenas and N Saffari

Ultrasonics Laboratory, Mechanical Engineering Department, University College London, London WC1E 7JE, UK

E-mail: seyyed.haqshenas.12@ucl.ac.uk

Abstract. Passive cavitation detectors are widely used for measuring acoustic emissions from cavitating bubbles. Acoustic emissions related to the dynamics of oscillating bubbles contain complex time and frequency domain information. Signal processing techniques traditionally used to analyse transient and stationary signals may be of limited value when analysing such acoustic emissions. This paper describes a multi-resolution approach developed for processing acoustic emissions data. The technique consists of the combination of a discrete wavelet transform and of the statistical and spectral analysis to extract cavitation features. These features include broadband emissions and harmonic, sub-harmonic and ultra-harmonic information. The implementation of the technique on experimental datasets demonstrates that this approach provides detailed information about key features of the acoustic signal, especially in complex situations where different types of cavitation occur simultaneously. Furthermore, statistical metrics used in this technique can provide a quantitative means for classifying signatures of cavitation, particularly the broadband segment of the spectrum created by inertial cavitation, which constitutes novel work.

1. Introduction

Acoustic cavitation is the formation and collapse of cavities (bubbles) in a liquid which undergoes a rapid pressure fluctuation generated by an acoustic source [1]. Reference is often made to two types of ultrasound induced cavitation, referred to as inertial and stable cavitation. The former refers to the situation where the created bubbles grow and implode violently within a few cycles of excitation, while in the case of stable cavitation, bubbles sustain and repetitively oscillate for a longer period of time [1]. Several physical effects are associated with each type of cavitation, namely the creation of free radicals, shock waves, shear forces, etc. [1]. These effects are exploited in a range of different applications including: non-invasive therapy [2], sono-chemistry [3], sono-crystallisation [4] and food processing [5]. In order to investigate and study the role of cavitating bubbles in these applications, the cavitation and bubbles oscillations have to be monitored and measured. Taking into account the aforementioned physical effects, several methods have been developed to detect and measure bubble activity [6]. Two main quantitative techniques for cavitation detection and measurement are passive cavitation detection [7] and active cavitation detection [8], which both exploit the acoustic emissions generated by cavitating bubbles. Passive techniques use a single element transducer, which is either focused or planar [9, 10], called a Passive Cavitation Detector (PCD), to capture acoustic emissions resulting from cavitation. Passive cavitation detection can be easily implemented and has been widely used over a range of applications [2, 11, 12].

Frequency analysis of the recorded time series acoustic emissions from cavitating bubbles has often been carried out by the aid of the Fast Fourier Transform (FFT) algorithm [13]. However, there exist a number of reasons as to why spectral analysis cannot thoroughly process readings of a PCD and can only extract limited information from such time series:

- cavitation is a transient phenomenon which only takes place over a short time period relative to that of the fundamental frequency. Transient events are inherently aperiodic, whereas the Fourier transform is best used for analysing signals containing periodic events [14];
- multiple phenomena occur simultaneously and appear in the spectrum as sub-harmonics, ultra-harmonics and a broadband segment, but they cannot be readily differentiated from each other and separately analysed through use of the Fourier transform alone;
- when one effect is greater in magnitude than another, e.g. strong ultra-harmonics but weak sub-harmonics, this may mask the signature of other frequency content in the spectrum and consequently affect its detectability; and
- spectral analysis only provides information about the frequency content of a signal but not the corresponding part of the signal in the time domain.

Another technique which is commonly used to analyse the acoustic emissions from cavitating bubbles relies on the combination of spectral analysis and digital filtering [2, 11, 12, 15]. Spectral analysis is performed to identify different components of the signal in the frequency domain. Digital filters are then designed accordingly to separate each component in the time domain. After temporally windowing the signal, each segment is filtered and represented by the value of its energy. These generated graphs are then used for visually assessing and characterising cavitation activity. This qualitative approach to identifying cavitation signatures also suffers from the shortcomings of spectral analysis. Additionally, it is only applicable to signals with a long exposure time, e.g. a few seconds, to allow for a sufficiently fine frequency resolution within each window.

One way to improve cavitation detection and classification in particular is to utilise Time-Frequency analysis techniques. Time-Frequency techniques are extensively used for analysing non-stationary signals whose frequency content and statistics change over time. Various methods have been developed for use on practical applications such as sonar, ultrasonic applications for non-destructive testing, fault diagnosis, remote sensing and many others. Short Time Fourier Transform (STFT), Bilinear (or quadratic) time-frequency distributions (also called Cohens Class distribution functions) and wavelet transforms, are tools developed for investigating both the temporal and the spectral information of a signal. The STFT is one of the time-frequency techniques most used in practice. It has also extensively been applied to acoustic cavitation detection [11, 15]. Perhaps the most novel signal analysis technique in this family is the wavelet transform, which was developed to address the shortcomings of the STFT [16]. Indeed, the main drawback of the STFT, especially in treating transient signals, is its fixed resolution: the broader the time window, the higher the frequency resolution, but the lower the time resolution, and vice versa. In contrast to the STFT, the wavelet transform is a multi-resolution transformation technique which gives good time resolution for high-frequency events and good frequency resolution for low-frequency events. In cavitation detection applications, the Discrete Wavelet Transform (DWT) has already been used for analysing hydraulic cavitation signals [17, 18]. In another application, the DWT was employed to characterise acoustic cavitation generated by an ultrasonic cleaning bath [19]. The wavelet transform has also been used for identification of contrast agent micro-bubbles [20]. In this paper, the DWT was combined with multi-scale spectrum analysis to develop a more powerful technique for acoustic cavitation detection and classification. Furthermore, statistical metrics were utilised for data evaluation and classification at each scale. This approach was implemented on signals obtained experimentally. The experimental protocol will be described in Section 3. Results demonstrate promising performance of the developed method in comparison with conventional techniques.

2. Materials and Methods

2.1. Signatures of cavitation in a PCD signal

Initially, it is necessary to identify signatures of inertial and stable cavitation in a PCD signal. Stable cavitation is highlighted by tonal components in the time series which are represented by harmonics (that is nf where $n = 2, 3, 4, \ldots$ and f is the excitation frequency) in the frequency domain. These harmonics are explained in terms of the forced nonlinear bubble oscillations though they can be due to nonlinear ultrasound wave propagation in the medium as well as nonlinear vibration of the ultrasound transducer, which can occur at high amplitude driving voltages. Inertial cavitation generates shock waves [21–23] which have impulsive characteristics in the time domain. This manifests itself as a broadband component of the signals spectrum [1, 12, 24]. Additionally, the spectrum of acoustic emission may contain sub-harmonics (that is f/n where $n = 2, 3, 4, \ldots$ and ultra-harmonics (that is mf/n where $m > n$ and m/n is non-integer). Several explanations have been given for these signatures though there is no overall agreement on them. One such explanation associates sub-harmonics with acoustic emissions from pulsating bubbles with equilibrium radii multiple times the size of the resonance radius [25–27]. Another study concluded that if bubbles with resonance radii are excited by harmonics of the fundamental frequency, sub-harmonics will appear in the spectrum [28–30]. Further work [31] showed that inertial forces, under circumstances where the pressure amplitude is larger than a specific threshold value, can delay the contraction phase of an oscillating bubble and therefore leading to bubble growth for a longer period. Consequently the oscillation frequency of the bubble is different from the excitation frequency. These features of the spectrum of a PCD signal can be summarised as follows:

- harmonics in the spectrum are an indication of oscillating bubbles, considered as stable cavitation;
- broadband noise is an indication of shock waves emitted by inertially cavitating bubbles; and
- ultra-harmonics and sub-harmonics can be assumed to be due to the oscillation of bubbles of either larger or smaller multiples of the resonant bubble size. Ultra- and sub-harmonics may also occurs when the contraction phase of the oscillating bubbles is delayed and consequently bubbles oscillate for more than one pressure cycle prior to collapse. There is no general agreement as to whether ultra- and sub-harmonics should be considered as an indication of inertial or stable cavitation though they are usually interpreted as a threshold for inertial cavitation.

These features can be identified in both time and frequency domains using appropriate statistical indices which will be further elaborated in the following sections.

2.2. Discrete wavelet transform

STFT essentially divides up a signal to different segments through windowing and applies an FFT to each segment. The choice of window function depends on the signal characteristics and its ability to suppress side-lobes and spectral leakage. STFT is mathematically defined as follows [32]:

$$S_x(t, f) = \int_{-\infty}^{\infty} x(t)\ w(t)\ e^{(-j2\pi ft)}\ dt \qquad (1)$$

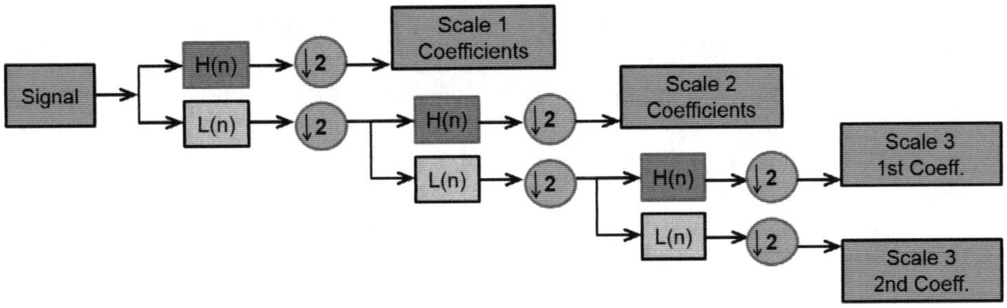

Figure 1: Mallat Pyramid algorithm, performing a DWT using a bank of filters determined by the wavelet function. $H(n)$ and $L(n)$ are transfer functions of highpass and lowpass filters. This figure shows a decomposition up to three scales.

where $w(t)$ is a window function, $x(t)$ is the signal in time domain, $x(t)w(t)$ is the windowed signal and f is the frequency. The Fourier transform uses the complex $e^{(j2\pi ft)}$ function as an analysing function to decompose a signal. In the case of wavelet transform, the analysing function is a wavelet function, ψ . The wavelet transform measures the correlation between signal and various wavelets which are created by shifting and scaling the mother wavelet ψ. This is mathematically written as follows [16]:

$$\mathcal{W}_x(u,s) \;=\; \int_{-\infty}^{\infty} x(t)\,\frac{1}{\sqrt{s}}\,\psi^*\left(\frac{t-u}{s}\right)\,dt \tag{2}$$

where $x(t)$ is the time signal, s and u are the scaling and translating parameters respectively and $*$ refers to the complex conjugate of the wavelet function. The continuous wavelet transform of a discrete signal can reveal an overwhelming amount of information buried in the signal, which may not be straightforward to interpret. Since in this form of the wavelet transform a continuous wavelet function is used, the scale and translation parameters are $u \in \mathbb{R}$ and $s \in \mathbb{R}^+$ respectively. Theoretically, these can assume infinite amount of values. This usually leads to redundancy as the spectrum changes very little between adjacent scales [16]. This sparked the enthusiasm for dicretization of the wavelet transform. The discrete wavelet transform is implemented using a bank of filters where the transfer functions of the filters are determined based on the mother wavelet function. The bandwidths of the highpass and lowpass filters, $H(n)$ and $L(n)$, at scale i are equal to $\left[\frac{f_s}{2^{i+1}}\ ,\ \frac{f_s}{2^i}\right]$ and $\left[0\ ,\ \frac{f_s}{2^{i+1}}\right]$, respectively, where f_s is the sampling frequency. This algorithm is illustrated in Figure 1. This method of performing the DWT with the aid of a bank of filters is the cornerstone of developing more complex signal processing techniques which would benefit from the multi-resolution analysis property of the wavelet transform. This algorithm was used to develop a signal processing technique which is suitable for the analysis of cavitation signals and is explained in the following section.

2.3. Multi-resolution cavitation analysis

The DWT algorithm was used to decompose PCD readings into a set of underlying components where each component has a particular frequency characteristic. This is equivalent to splitting up the signal into its basic elements over different frequency ranges. Summing these individual components results in the whole signal being reconstructed. Each component was then analysed separately to extract cavitation features. The simplest processing method is to perform a spectral analysis of each component and to look for aforementioned signatures of cavitation. The energy of each component can also be calculated as an estimate of the intensity of physical phenomena, i.e. the type of cavitation. In order to pinpoint different regimes of cavitation quantitatively, two statistical indices were employed for feature extraction. Since shock waves created by the inertial cavitation of bubbles appear as short impulses in the time domain, one expects more "peaky" signal when inertial cavitation occurs. On the other hand, the time series would contain more tonal elements which would exist for longer time periods when bubbles oscillate in a stable fashion. Consequently, the signal would be less impulsive. These indices were kurtosis and crest factor which are explained below.

Kurtosis: Assuming a signal is a realisation of a real random variable, kurtosis then measures the "peakedness" of the probability distribution of this random variable. This can be interpreted as the proportion of the variance which is the result of infrequent extreme deviations. The mathematical formulation of kurtosis is as follows [33]:

$$\mathbf{K} = \frac{\mathbf{E}(x - \mu)^4}{\sigma^4} \tag{3}$$

where $\mathbf{E}()$ is the expected value operator, x is a random variable (signal) and μ and σ are the mean value and standard deviation of x respectively.

Crest factor: Crest factor is the ratio of peak values of a signal to the average value of the signal. The crest factor may be interpreted as a measure of how "extreme" the peaks in the signal are. A higher crest factor tends to indicate more impulsive behaviour of the signal. The mathematical formulation of the crest factor expressed in decibels is as follows [34]:

$$\mathbf{C}_f = 10 \log \frac{\max(|x|)^2}{x_{rms}^2} \tag{4}$$

where $|x|$ is the absolute value of variable x. The crest factor is more sensitive to small and short impulses compared to the kurtosis. The kurtosis and crest factor of a pure tone can be found to be about 1.5 and 3 dB, respectively. In the case of a Gaussian noise, the kurtosis is about 3 whereas the crest factor can be any real number as the Gaussian distribution is theoretically unbounded. Given these values, a threshold value for classifying cavitation regimes, however, cannot be readily defined. Nonetheless, these statistical metrics can be utilised for the comparative study of contents of different scales of signal decomposed by the DWT.

The flow chart shown in Figure 2 demonstrates the signal processing technique employed in this study. Taking into account the frequency response of a PCD transducer, those scales of decomposition which fall in the frequency bandwidth of the transducer should be acquired and analysed.

3. Experimental Results and Discussion

A schematic of the test set-up is depicted in Figure 3 and briefly explained in Table 1. The transducers were immersed in a tank filled with de-ionised water at room temperature (22°C). The gas content of water was measured by an oxygen meter probe and kept about 7.00±0.2 mg/L for all trials. The excitation signal was a burst of 100 cycles with a fundamental frequency of 1.1 MHz. Tests were carried out under different excitation amplitudes, the values of which are displayed in Table 2. This table also includes the estimated peak positive and negative pressure amplitude at the focus under these excitation conditions. These estimations were obtained by numerically solving the KZK (Khokhlov-Zabolotskaya-Kuznetsov) equation in MATLAB using the HIFU simulator software [35]. The KZK equation describes the propagation of

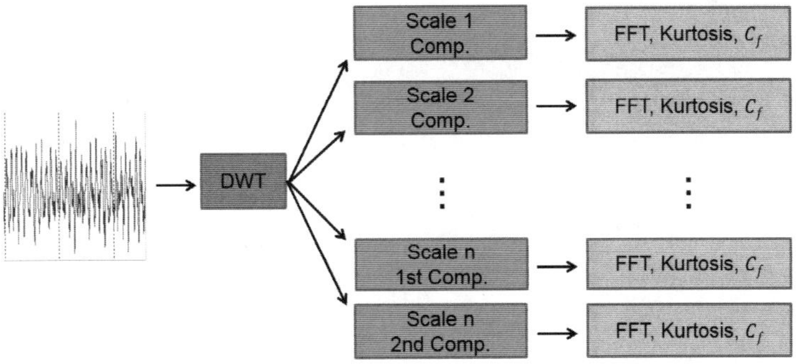

Figure 2: Algorithm for the multi-resolution of a cavitation signal.

acoustic waves and accounts for effects due to nonlinear wave propagation, diffraction and absorption [36]. The conditions that under which the KZK equation is valid are expressed as follows [37]: i) $d/a > 1$ where d is the active diameter of a focused transducer and a is the focal depth; and ii) $ka > (d/a)^{\frac{1}{3}}$ where k is the wavenumber. The geometry of the HIFU transducer used in this work together with the large wavenumber, i.e. 4600 m^{-1}, satisfy both of these conditions. Acoustic emissions from bubbles at the focus were picked up by two focused transducers which were confocal with the HIFU transducer. These receivers are depicted as R1 and R2 in Figure 3. The acquired voltage signals from the receiving transducers were amplified and digitised at the sampling rate of 200 MHz using an oscilloscope. The effect of the transfer function of each transducer was corrected by performing a deconvolution using digital filters implemented in the time domain. Having measured the pulse-echo transfer function of transducers, deconvolution filters were determined using a least squares algorithm which employs derivative regularization [38]. The Daubechies wavelet function of order 10 [16] was selected and eight scales of decomposition were chosen for the signal processing algorithm. The type of wavelet function was selected in accordance with the physics of cavitation and impulsive nature of desired component, as discussed in Section 2.1. The depth (i.e. the number of scales) of the wavelet transform was chosen by considering the receiving transducers' bandwidth and the sampling rate used in the measurements. This should be noted that the type of wavelet function and depth of decomposition which are best adapted to process the data are not unique. Indeed, other orthonormal wavelet functions with relatively high order should be able to extract cavitation signatures from the signal.

The spectra of pre-processed signals measured by R1 and R2 transducers are depicted in Figure 4. Figures 5 to 7 show the three underlying components of readings from the second transducer, i.e. within the 0.4 MHz to 3.2 MHz bandwidth, and their corresponding spectra for test condition two. Sub-harmonics are clearly detectable in Figure 5. The crest factor and kurtosis of this component of the signal are 12 dB and 4 respectively, which suggests the existence of an impulsive component in this time series. This is also manifested through a low amplitude erratic segment of the components spectrum, from 0.2 MHz to 0.85 MHz. Figure 6 shows the component of the signal which is merely comprised of the fundamental tone, i.e. the excitation frequency. The very low crest factor and kurtosis also confirm the tonal nature of this component. Figure 7 shows the component of the signal in the frequency range of 1.6 MHz to 3.12 MHz, together with its spectrum. This component consists of the second harmonic of the driving signal, and a set of elements with broadband characteristics. The high crest factor value of 16 dB, together with high kurtosis value of 7.5, suggests an impulsive nature of this component of the signal, which is also reflected in the broadband element of the spectrum.

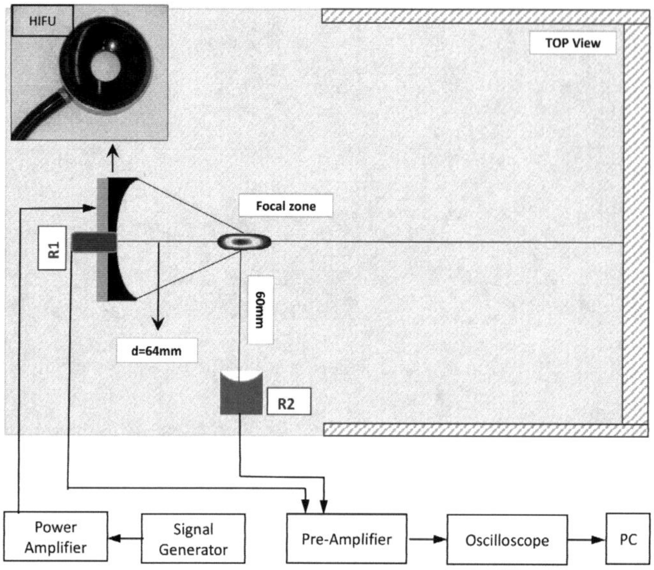

Figure 3: Schematic test setup. Components are described in Table 1.

Table 1: Specification of the hardware of test setup

Component	**Type**	**Specification**
HIFU	Soniconcept H102	High Intensity Focused Ultrasound transducer, $f_0 = 1.1$ MHz, -3 dB focal width is 1.33 mm by 13.5 mm
R1	Olympus V327 type	focused transducer with centre frequency of 10 MHz, -6 dB bandwidth is about 8 MHz
R2	Olympus V392 type	focused transducer with centre frequency of 1 MHz, -6 dB bandwidth is about 1.2 MHz
Power Amplifier	*E&I* 1040L type	class AB RF amplifier with nominal gain of 55 dB
Pre-Amplifier	Stanford Research Systems, SR445A	linear preamplifier with bandwidth of 350 MHz, gain of 7 dB
Oscilloscope	LeCroy HRO64Zi	two channels were used simultaneously for recording data, sampling rate of 200 MHz per channel
Oxygen meter	VWR OX4000 Set	the instrument measures gas content and temperature of water

Table 2: Test condition

Test condition	**1**	**2**	**3**	**4**
Driving voltage (mVpp)	80	120	200	280
P_-(MPa)*	2.6	3.5	5.8	7.1
P_+(MPa)*	3	4.5	8.7	12.5

* this is an estimated value using the KZK equation solver.

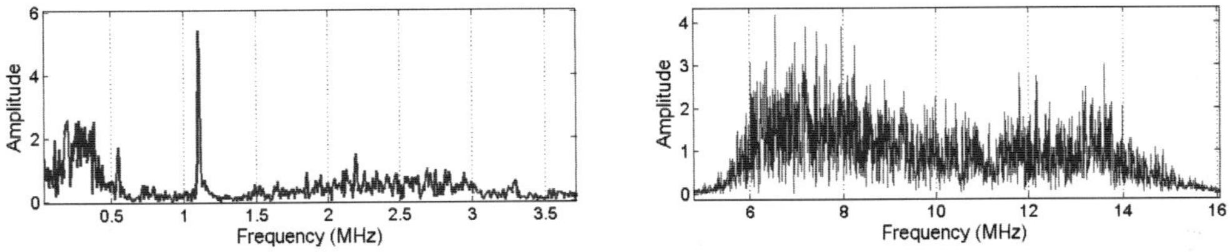

Figure 4: Spectra of signal measured by: left: the 1 MHz transducer, right: the 10 MHz transducer.

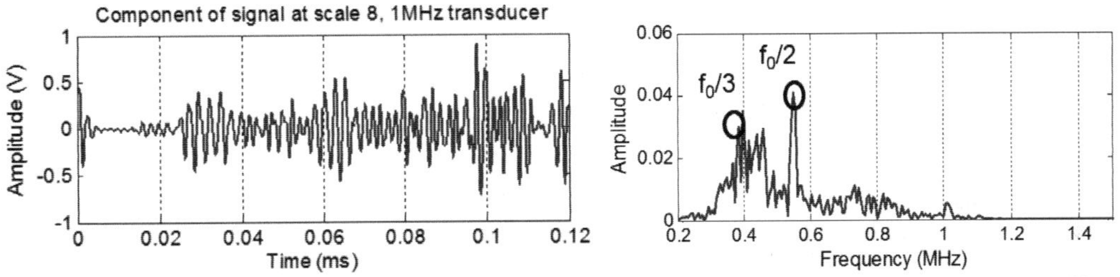

Figure 5: Component of the signal measured by the 1 MHz transducer at scale 8 (frequency range of 0.4 MHz to 0.8 MHz) and its spectrum. $C_f = 12$ dB, $K = 4$.

The high value of the statistical metrics along with the broadband element in the spectrum, highlights the probability of occurrence of inertial cavitation which is present in this component of signal. It is difficult to detect the broadband part of the signal in the frequency range of 1.5 MHz to 3 MHz, using conventional Time-Frequency techniques. Indeed the broadband component can easily be masked by the fundamental and/or its harmonics if these are of high enough amplitude. Figure 8 shows the component of the signal picked up by the transducer R1. This component has a very high crest factor, in this case 17 dB and consequently, is likely to contain a broadband spectrum, which is observable in the figure. The high values of the statistical metrics, e.g. kurtosis above 3, indicates the presence of a broadband component in the signal which is produced by the impulsive acoustic emissions of inertial cavitation occurring at the focus. Low values of these metrics, however, do not necessarily refute the possibility of inertial cavitation as the sub-harmonics and ultra-harmonics signatures of inertial cavitation should be analysed too. Another observation which can be made is the existence of multiple broadband segments, with different energy levels, which are difficult to identify by performing a spectral analysis of data on a single scale, i.e. through use of an FFT applied to the time-series.

This analysis was repeated on other data sets acquired under the different test conditions de-

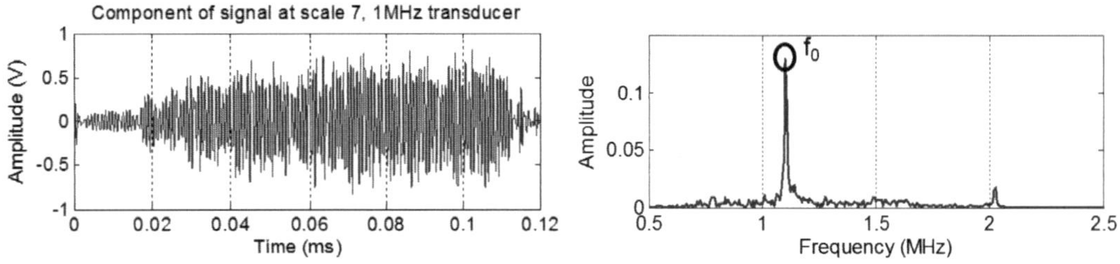

Figure 6: Component of the signal measured by the 1 MHz transducer at scale 7 (frequency range of 0.8 MHz to 1.6 MHz) and its spectrum. $C_f = 8$ dB, $K = 2.3$.

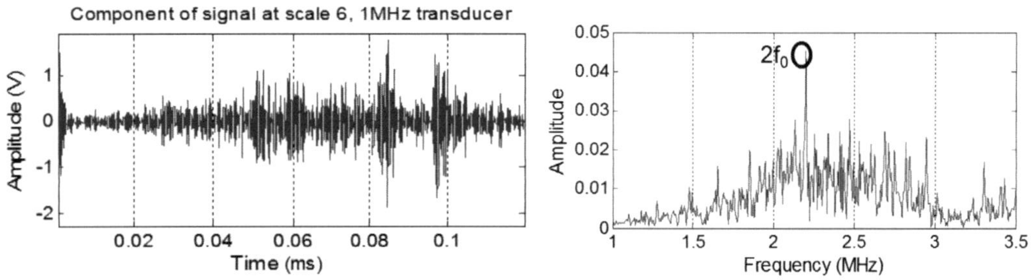

Figure 7: Component of the signal measured by the 1 MHz transducer at scale 6 (frequency range of 1.6 MHz to 3.12 MHz) and its spectrum, $C_f = 16$ dB, $K = 7.5$.

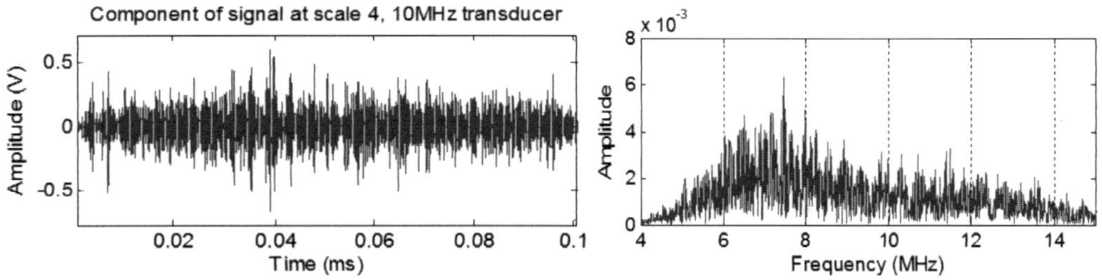

Figure 8: Component of the signal measured by the 10 MHz transducer at scale 4 (frequency range of 6.25 MHz to 12.5 MHz) and its spectrum, $C_f = 17$ dB, $K = 4.7$.

scribed in Table 2. Furthermore, the energy of each sub-harmonic component, ultra-harmonics and broadband features was calculated, as these components have already been differentiated by the aid of the multi-resolution technique developed in this paper. As explained in the Section 2.2, the bandwidth of filters of the DWT varies as a dyadic function of the scale number. At low scale numbers, the bandwidth is broad and it gets narrower as the scale number becomes larger. Consequently, it is possible that in a single scale, particularly in a low scale number, two different spectral components occur simultaneously. In the case of overlap of a broadband component with harmonics or ultra-harmonics, detected using statistical indices, a notch filter was designed to remove tonal components. This enabled the energy of each component to be estimated separately. The bar charts depicted in Figure 9 display the magnitude of the energy, in arbitrary unit, of sub-harmonics and broadband components under all four test conditions. The variance of time domain components were calculated as the measure of the energy of the signal. The results in Figure 9 demonstrate that the raising driving voltage and creating an acoustic

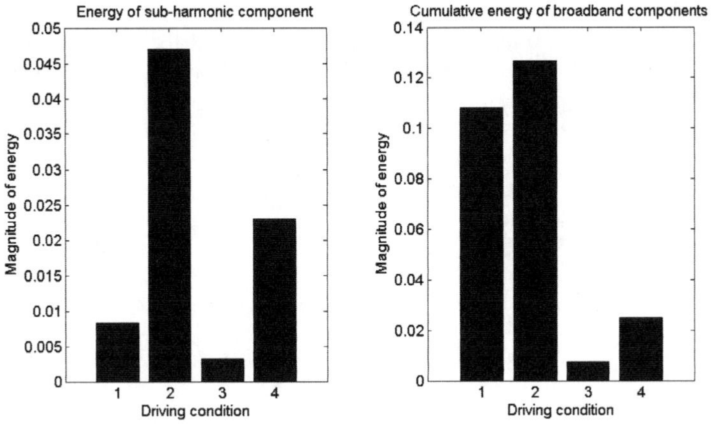

Figure 9: Comparing energy of different components at different test conditions.

pressure field of larger magnitude at the focus (see Table 2 for estimated pressure magnitudes at each condition), does not necessarily enhance the broadband characteristic of inertial cavitation resulting from shock waves. Moreover, one can see that at a constant driving frequency, tuning the excitation amplitude not only yields different cavitation regimes, which is expected from theory, but also favours or suppresses different features (or signatures) of the cavitation regime. The latter effects have been numerically demonstrated by Lauterborn [28]. The multi-resolution technique shows promise in being capable of monitoring and quantifying this behaviour using experimental data.

4. Conclusion

A new technique for the application of acoustic cavitation detection and characterisation was described in this paper. The technique utilises multi-resolution analysis together with statistical and spectral analysis of acoustic emissions radiated by cavitating bubbles. The technique uses the discrete wavelet transform to decompose a signal into its underlying components and performs the statistical and spectral analysis on each component to extract cavitation features. These features include broadband emissions and harmonic, sub-harmonic and ultra-harmonic information. It was demonstrated that in complex cases where both types of cavitation occur simultaneously with overlapping spectral features, this technique can identify and quantify each of these successfully. This capability is provided by using statistical metrics and spectral analysis at each scale of the wavelet transform. Moreover, thanks to the multi-resolution property of the DWT, this technique can be applied to signals which are impulsive in nature. This makes the multi-resolution method a useful tool for studying cavitation in short exposure shots. In spite of the improvements in processing acoustic emissions sensed by PCDs aiming for cavitation detection and characterisation, this technique is likely to be further enhanced by using more flexible wavelet transform approaches and feature extraction techniques. Further work is currently in progress which aims to investigate the use of wavelet packet transforms for performing multi-resolution analysis and incorporating more advanced statistical techniques for feature extraction and classification.

Acknowledgments
The authors would like to acknowledge and thank A. Gavriilidis for his advice on crystallisation and P. Gelat for thoughtful discussions. This work was made possible through funding provided by the EPSRC (grant EP/I031480/1).

References
[1] Leighton T 1997 *The acoustic bubble* (Academic Press) ISBN 0124419208
[2] Coussios C C and Roy R A 2008 *Annual Review of Fluid Mechanics* **40** 395–420
[3] Peuker U A, Hoffmann U, Wietelmann U, Bandelin S and Jung R 2006 *Ullmann's Encyclopedia of Industrial Chemistry*
[4] Ruecroft G, Hipkiss D, Ly T, Maxted N and Cains P 2005 *Organic Process Research & Development* **9** 923–932
[5] Povey M 1997 *Ultrasonic techniques for fluids characterization* (Academic Press) ISBN 9780125637305
[6] ANSI 2007 S1 24 T 2002 *ANSI technical report, Bubble detection and cavitation monitoring* (American National Standard Institute)
[7] Atchley A, Frizzell L, Apfel R, Holland C, Madanshetty S and Roy R 1988 *Ultrasonics* **26** 280 – 285 ISSN 0041-624X
[8] Roy R A, Madanshetty S I and Apfel R E 1990 *The Journal of the Acoustical Society of America* **87** 2451
[9] Madanshetty S I, Roy R A and Apfel R E 1991 *The Journal of the Acoustical Society of America* **90** 1515–1526
[10] Mast T D, Salgaonkar V A, Karunakaran C, Besse J A, Datta S and Holland C K 2008 *Ultrasound in Medicine & Biology* **34** 1434–1448
[11] McLaughlan J, Rivens I, Leighton T and ter Haar G 2010 Commentary on the detection of bubble activity generated in ex-vivo tissue by high intensity focused ultrasound (hifu) with respect to the generation of therapeutic lesions in tissue for the treatment of cancer ISVR Technical Report 330 University of Southampton
[12] McLaughlan J, Rivens I, Leighton T and ter Haar G 2010 *Ultrasound in Medicine & Biology* **36** 1327–1344
[13] Cramer E and Lauterborn W 1982 *Applied Scientific Research* **38** 209–214
[14] Adams M 2012 *Continuous-time signals and systems* (Michael Adams) ISBN 9780987919700
[15] Farny C H, Holt R G and Roy R A 2009 *Ultrasound in Medicine & Biology* **35** 603–615
[16] Mallat S 1999 *A Wavelet tour of signal processing* Wavelet Tour of Signal Processing (Elsevier Science) ISBN 9780080520834
[17] Shi G Z and Hu J C 2007 Ship noise demodulation line spectrum fusion feature extraction based on the wavelet packet *International Conference on Wavelet Analysis and Pattern Recognition.* vol 2 (IEEE) pp 846–850
[18] He Y and Liu Y 2011 *Applied Acoustics* **72** 721–731
[19] Moholkar V, Huitema M, Rekveld S and Warmoeskerken M 2002 *Chemical Engineering Science* **57** 617–629
[20] Zong Y, Li B, Wan M and Wang S 2006 Detection of individual microbubbles using wavelet transform based on a theoretical bubble oscillation model *Advances in Natural Computation* (Springer) pp 354–363
[21] Matula T J, Hallaj I M, Cleveland R O, Crum L A, Moss W C and Roy R A 1998 *The Journal of the Acoustical Society of America* **103** 1377–1382
[22] Matula T J 1999 *Philosophical Transactions of the Royal Society of London, Series A: Mathematical, Physical and Engineering Sciences.* **357** 225–249
[23] Pecha R and Gompf B 2000 *Physical review letters* **84** 1328
[24] Walton A J and Reynolds G T 1984 *Advances in Physics* **33** 595–660
[25] Neppiras E 1969 *The Journal of the Acoustical Society of America* **46** 587–601
[26] Neppiras E 1969 *Journal of Sound and Vibration* **10** 176–186
[27] Eller A and Flynn H 1969 *The Journal of the Acoustical Society of America* **46** 722–727
[28] Lauterborn W 1976 *The Journal of the Acoustical Society of America* **59** 283–293
[29] Lauterborn W and Parlitz U 1988 *The Journal of the Acoustical Society of America* **84** 1975–1993
[30] Prosperetti A 1975 *The Journal of the Acoustical Society of America* **57** 810–821
[31] Akulichev V 1971 Pulsations of cavitation voids *High-Intensity Ultrasonic Fields* (Springer) pp 201–259
[32] Oppenheim A V and Schafer R *Digital signal processing* ISBN 9780132146357
[33] Vaseghi S V 2008 *Advanced digital signal processing and noise reduction* (John Wiley & Sons)
[34] Norton M and Karczub D *Fundamentals of noise and vibration analysis for engineers* ISBN 9780521499132
[35] Soneson J 2011 Hifu simulator matlab package v1.2
[36] Hamilton M and Blackstock D 1998 *Nonlinear Acoustics* (Academic Press) ISBN 9780123218605
[37] Soneson J E 2012 *The Journal of the Acoustical Society of America* **131** EL481–EL486

[38] Mandersson B and Salomonsson G 1989 *IEEE Transactions on Ultrasonics, Ferroelectrics and Frequency Control.* **36** 109–113

Journal of Physics: Conference Series **581** (2015) 012005 doi:10.1088/1742-6596/581/1/012005

An optimised stiffness reduction method for simulating infinite elastic space using commercial Finite Elements codes

J R Pettit[1, 2], A Walker[1] and M J S Lowe[2]

[1] Rolls-Royce Nuclear, PO BOX 2000, Derby, UK, DE21 7XX

[2] Research Centre for NDE, Imperial College London, Exhibition Road, London, UK, SW7 2AZ

James.Pettit@Rolls-Royce.com

Abstract. A common goal when using Finite Element (FE) modelling in time domain wave scattering problems is to minimise model size by only considering a region immediately surrounding a scatterer or feature of interest. The model boundaries must simulate infinite space by minimising the reflection of incident waves. This is a significant and long-standing challenge that has only achieved partial success. Industrial companies wishing to perform such modelling are keen to use established commercial FE packages that offer a thorough history of validation and testing. Unfortunately, this limits the flexibility available to modellers preventing the use of popular research tools such as Perfectly Matched Layers (PML). Unlike PML, Absorbing Layers by Increasing Damping (ALID) have proven successful offering practical implementation into any solver that has representation of material damping. Despite good performance further improvements are desirable. Here, a Stiffness Reduction Method (SRM) has been developed and optimised to operate within a significantly reduced spatial domain. The technique is applied by altering damping and stiffness matrices, inducing decay of incident waves. Variables are expressed as a function of known model constants, easing implementation for generic problems. Analytical and numerical solutions have shown that SRM out performs ALID, with results approaching those of PML.

1. Introduction

Commercial Finite Element (FE) codes have been used increasingly to simulate elastic wave propagation and scattering problems [1, 2]. Numerical techniques can offer solutions that extend to any desired level of complexity. However, the greatest limitation when using FE methods is the computational resource that can be allocated to solving the problem. To maintain a highly accurate solution, novel techniques have been developed that allow for the use of reduced spatial domains that do not compromise on having an accurate representation of the system.

Such methods include the use of domain linking algorithms [3-5] and the simulation of infinite elastic space. These methods allow modelers to only consider a small area immediately surrounding a scatterer or feature of interest. Computational resource can therefore be focused on a highly accurate representation of the Area of Study (AoS). This is practically advantageous when considering three-dimensional problems where even a significantly reduced spatial domain can consist of tens of millions of degrees of freedom to be solved.

When using such techniques, a common problem is the removal of unwanted reflections from model boundaries manifesting themselves as additional remnants in ultrasonic scattering signals. The removal of unwanted reflections has proven to be a challenging problem. Innovative techniques have been developed which can be collated into the three categories: infinite element methods [6], non-reflecting boundary conditions [7], and absorbing layer methods [8-11].

Despite the advances made with these methods, they do not meet all of the bespoke requirements that are necessary for industrial Non-Destructive Evaluation (NDE) applications using commercial FE codes. The relative high performance and ease of implementation of absorbing layer methods has meant that they have become the most viable tool for this application. Two successful absorbing layer techniques are Absorbing Layers by Increasing Damping (ALID) [8], and Perfectly Matched Layers (PML) [9].

Rajagopal et al [11], performed a comparative study of absorbing layer methods and showed that PML generally achieves a smaller model size than ALID. However, PML is more suited to frequency domain simulations and requires specific programming and functionality; currently this is only available in a minority of commercial FE packages. Therefore, ALID is a more accessible tool for practical use, being directly compatible with explicit time domain solvers in commercial FE codes.

Despite the good performance of ALID, further improvements are desirable. The Stiffness Reduction Method (SRM) is an extension on current absorbing layer techniques that improves upon the efficiency of ALID, whilst still remaining accessible to implementation using commercial FE programs. The method which will be presented in the following sections, exploits benefits that can be gained by making gradual changes to the material stiffness, as well as increasing the damping, across the absorbing region.

2. Theory of SRM

We start by reviewing the underlying workings of absorbing layer methods, from which we will be able to identify the role of the stiffness term and thus develop the new stiffness reduction approach. A full derivation of this approach is given by Pettit et al [12].

The absorbing layer concept changes the physical properties of the elements within successive layers such that they absorb incident wave energy. The physical properties must be changed gradually, but at an optimum rate because of two competing mechanisms.

The first is that reflections are caused by impedance mismatches between successive absorbing layers. Thus, alterations in the physical properties of adjacent elements cause reflections that return energy back into the AoS at each interface.

The second is that the incident wave may not be fully absorbed within the total thickness of the absorbing region and reflect from the model boundary back into the AoS. An efficient absorbing boundary balances these two mechanisms to produce a net reflection below a pre-defined tolerable threshold from a minimal thickness of absorbing material.

The equation of dynamic equilibrium that forms the basis of the FE discretisation of the wave propagation problem is:

$$[F] = [M]\ddot{u} + [C]\dot{u} + [K]u \tag{1}$$

where [M], [C] and [K] are the respective mass, damping and stiffness matrices and [F] represents external force.

The damping matrix can be expressed as two separate components proportional to the mass and stiffness matrices, consistent with most commercial FE packages:

$$[C] = C_M[M] + C_K[K] \tag{2}$$

Combining equation (1) and equation (2) produces a new equation for dynamic equilibrium where damping is expressed solely as a function of coefficients C_M, mass proportional damping and C_K, stiffness proportional damping:

$$-[M]\left(1+\frac{iC_M}{\omega}\right)\omega^2 u + [K](1-i\omega C_K)u = [F] \tag{3}$$

From the inspection of the coefficients of equation (3) it can be shown that instead of the damping matrix [C], there are complex values of the density, ρ, and Young's modulus, E, that can be expressed as:

$$\rho \rightarrow \rho\left(1+\frac{iC_M}{\omega}\right) \tag{4}$$

and

$$E \rightarrow E(1-i\omega C_K) \tag{5}$$

Inside the AoS, damping is usually zero ($C_M = 0$ and $C_K = 0$). Inside the absorbing region, damping is non-zero, meaning that incident wave energy is absorbed. This is attributed to the imaginary components of terms introduced into the solution via ρ and E. This can be illustrated by noticing that the wavenumber, k, is related to the density and stiffness such that:

$$k \propto \sqrt{\frac{\rho}{E}} \tag{6}$$

which becomes

$$k(k_{real}, k_{imag}) \propto \sqrt{\frac{\rho\left(1+\frac{iC_M}{\omega}\right)}{E(1-i\omega C_K)}} \tag{7}$$

Inside the absorbing layers the wavenumber has both real and imaginary parts such that:

$$k(k_{real}, k_{imag}) = k_{real} + ik_{imag} \tag{8}$$

Substituting this into the standard harmonic solution reveals an exponential decay term, $exp[-k_{imag}x]$, that attenuates waves within the absorbing region.

$$u(x,t) = exp[i(k_{real}x - \omega t)]exp[-k_{imag}x] \tag{9}$$

It is the imaginary component of the wavenumber which induces the decay of the waves inside the absorbing region and therefore that must be exploited. Commercial FE packages do not offer a means to alter the wavenumber directly. However, this can be achieved indirectly, by altering values of the stiffness, damping and mass matrices, to increase the value of k_{imag} inside the absorbing region. Consequently, this will also increase the value of k_{real} which must be managed carefully in order to minimise the reflection between adjacent elements.

2.1. Absorbing layer properties

Layer properties must be carefully selected such that successive values of k_{real} result in impedance mismatches that generate reflections which can be considered negligible. Progress made in ALID by Rajagopal et al [11], gives a good indication of how best to optimise initial parameters and forms the basis of the SRM technique.

Firstly, it is noted that introducing damping into the model can decrease the value of the stable time increment within explicit schemes, thereby reducing computational efficiency. However, this effect is only noticeable when using C_K. Setting C_K to zero eliminates this issue and allows damping to be controlled solely by C_M in which case it is found that the stability is not compromised. The use of absorbing layer methods with a non-zero C_K is discussed by Semblat et al [10], relating to the Rayleigh/Caughey damping formulation.

Secondly, damping is defined as an increasing power law set by the power p, so as to gradually alter the value across successive layers.

$$C_M(x) = C_{Mmax}X(x)^p \tag{10}$$

C_{Mmax} is a positive real number, where X(x) ranges from 0 at the AoS boundary to 1 at the end of the absorbing region. The thickness of each absorbing layer is equal to one element width, dx.

From inspection of equation (10), it can be seen that the imaginary component of the wavenumber, k_{imag}, can be incrementally increased by using $C_M(x)$ but also by decreasing the value of the Young's Modulus, E.

$$k(k_{real}, k_{imag}) \propto \sqrt{\frac{\rho}{E} + \frac{iC_M\rho}{\omega E}} \tag{11}$$

Therefore within the absorbing region E becomes a function of x, decreasing gradually across the absorbing region, avoiding any dramatic changes in material properties. Equation (11) indicates how best to increment variables between successive layers by using an exponential decay function containing an attenuation factor such that:

$$E(x) = E_0 exp[-\propto(x)k_{inc}x] \tag{12}$$

where E_0 is the Young's Modulus inside the AoS and k_{inc} is the incident wavenumber. However, the attenuation factor $\alpha(x)$ cannot be a constant for each layer since this will induce impedance mismatches between adjacent layers. Therefore, it is necessary to gradually increase this value across the absorbing region as shown in equation (13).

$$\propto(x) = \propto_{max} X(x)^p \tag{13}$$

The SRM still maintains the mass proportional damping $C_M(x)$, which is varied in the same manner as ALID, but now includes the additional reductions made to the stiffness matrix. The combination of these two effects allows for the attenuation of incident waves within a significantly reduced spatial domain.

2.2. *Defining absorbing layer input parameters*

Having derived functions to vary the stiffness and damping matrices, it is now necessary to calculate input parameters that define the absorbing region. The procedure outlined here is transferable to other absorbing boundary methods. The influence these variables have on attenuating an incident wave is dependent upon the angular frequency, ω, and FE mesh density, N, which is given by:

$$N = \frac{2\pi}{k_{inc}dx} \tag{14}$$

From inspection of equation (11), it can be seen that $C_M(x)$ is intrinsically linked to the angular frequency of the incident wavelength, ω. From trial and error methods carried out in this study, it has been observed that the optimal value selected for C_{Mmax} is of the same order as ω.

This makes the application of the absorbing regions to different inspection scenarios more straightforward, since now variables can be expressed as a function of the frequency of the incident wave rather than resulting to trial and error methods.

The power to which successive layers are raised, p, is best expressed as a cubic variation so that the interface between layers is continuous to the 1st and 2nd order. This is consistent with findings from Rajagopal et al [11], who stated that this is sufficient to minimise impedance mismatches between successive layers. The larger impedance mismatches that do occur happen deep within the absorbing region.

At this point the input parameters for SRM and ALID differ. For the SRM, it is necessary to define the maximum attenuation of the stiffness matrix, α_{max}, in the final absorbing layer where x=L. This is done by rearranging equation (12) to make $\alpha(x)$ the subject of the equation and defining a new variable α_{SRM}; the ratio of the Young's modulus in the final absorbing layer and the AoS.

$$\alpha_{max} = \alpha(L) = \frac{ln(\alpha_{SRM})}{k_{inc}L} \tag{15}$$

where

$$\alpha_{SRM} = \frac{E(L)}{E_0} = 0.01 \tag{16}$$

Just as a generic value of C_{Mmax} has been chosen through inspection and best practice, so too has α_{SRM}. It has been observed that setting the Young's Modulus in the final absorbing layer to be 1% of the value in the AoS will render a good performing absorbing layer. When using ALID, the stiffness is not varied, such that $\alpha_{SRM} = 1$.

Because the SRM can dampen out signals more readily than ALID, the SRM can operate in a reduced spatial domain. Work carried out by Rajagopal et al [11], indicates that an optimal thickness of ALID would be equal to $3\lambda_{inc}$. Due to the improved performance of SRM, that same level of performance can be achieved within a thickness of $1.5\lambda_{inc}$.

3. Comparison of SRM and ALID

To quantify the performance of a set of absorbing layers, it is necessary to calculate the magnitude of unwanted reflections that return from the model boundary. This is achieved through the use of an analytical approach based upon the Global Matrix method [13]. The analytical model is then validated by making comparisons from a FE model that contains the same absorbing boundary. The package ABAQUS/Explicit [14], has been chosen for the study, as it is widely available, developed and supported for wave propagation simulations.

The analytical and numerical models are non-dimensional and the material properties are defined such that the longitudinal and shear wavelengths are 1 and 2 respectively; each incident wave mode is considered separately.

To illustrate the advantages offered by the SRM, the performance is compared to conventional ALID. For both cases, the length of the absorbing region is equal to 1.5 times the incident wavelength, λ_{inc}. The attenuation parameters are defined such that for the SRM, the $\alpha_{SRM} = 0.01$, $C_{Mmax} = \omega$ and $p = 3$. For ALID, the attenuation parameters are defined according to work presented by Rajagopal et al [11], where the stiffness is not reduced and $C_{Mmax} = 5\omega$.

3.1. Analytical model

The Global Matrix method establishes a matrix relationship which describes plane wave propagation across all the layers of a multi-layered system. Each layer contains a local description of the wave, where adjacent layers are coupled by applying Snell's law for continuity of stress and displacement.

Rajagopal et al [11], successfully applied this technique to analyse the performance of the ALID by gradually implementing damping using the complex expression for density. The extension made here is to include the reduction in Young's modulus, as well as the complex density used by ALID.

The complete global matrix for waves propagating into an absorbing region is illustrated in figure 1 and defined by equation (17).

$$\begin{bmatrix} [M_1^b] & [M_2^f] & 0 & \cdots & 0 \\ 0 & [M_2^b] & [M_3^f] & \cdots & 0 \\ \cdots & \cdots & \cdots & \cdots & 0 \\ 0 & 0 & 0 & [M_{end-1}^b] & [M_{end}^f] \\ 0 & 0 & 0 & 0 & [M_{end}^b] \end{bmatrix} \begin{bmatrix} [neg_1] \\ \begin{bmatrix} pos_2 \\ neg_2 \end{bmatrix} \\ \vdots \\ \begin{bmatrix} pos_{end-1} \\ neg_{end-1} \end{bmatrix} \\ \begin{bmatrix} pos_{end} \\ neg_{end} \end{bmatrix} \end{bmatrix} = \begin{bmatrix} [M_{pos_1}^b] \quad [pos_1] \\ 0 \\ \vdots \\ 0 \\ 0 \end{bmatrix} \tag{17}$$

where $[M^f_n]$ and $[M^b_n]$ are the coefficients defined by the material properties of each layer, n, governed by Hooke's law and the Helmholtz decomposition linking displacement, u, with the compression and shear wave potentials, at the front boundary, f, and back, b boundary in the positive, pos, and negative, neg, propagation directions, with the subscript 'end' denoting the final absorbing layer.

For a given model of known material properties the left hand matrix is known. Similarly, for a specified incident wave the initial amplitude propagating into the absorbing region [pos_1] is also known. Matrix inversion is used to calculate the value of the wave amplitudes within each layer, thereby calculating the amplitude of the wave returning into the AoS, [neg_1]. Since the amplitude of the incident wave mode is known, this gives the reflection coefficient of both reflected wave modes and for angles of incidence within the range 0° to 90 °.

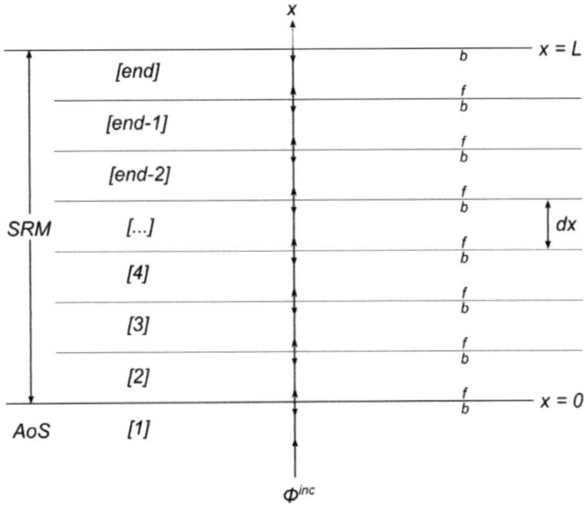

Figure 1. Figure to show application of the Global Matrix method to an absorbing boundary for an incident compression wave, φ^{inc}, entering the SRM from the AoS.

3.2. Numerical model

A harmonic bulk wave is incident on an absorbing region across a range of incident angles θ_{inc}, using FE code ABAQUS/Explicit [14], see figure 2.

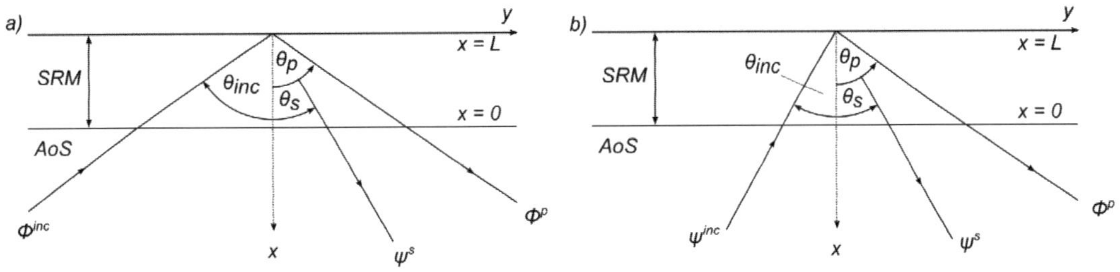

Figure 2. FE model of absorbing boundary for a) an incident compression wave, φ^{inc}, and b) and incident shear wave, ψ^{inc}, entering an absorbing region of thickness L, from the AoS.

The amplitudes of the incident and reflected wave modes are monitored along their predicted lines of propagation and the reflection coefficient deduced.

Results for the analytical and numerical models are shown in figure 3 and figure 4 for an incident longitudinal wave and in figure 5 and figure 6 for an incident shear wave.

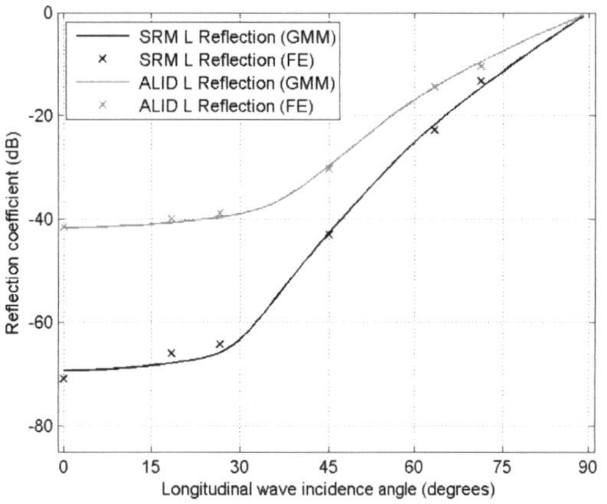

Figure 3. Global Matrix Method (GMM) and numerical (FE) reflection coefficients as a function of incident angle for the reflected compression wave mode (L) from SRM and ALID absorbing boundaries for an incident longitudinal bulk wave.

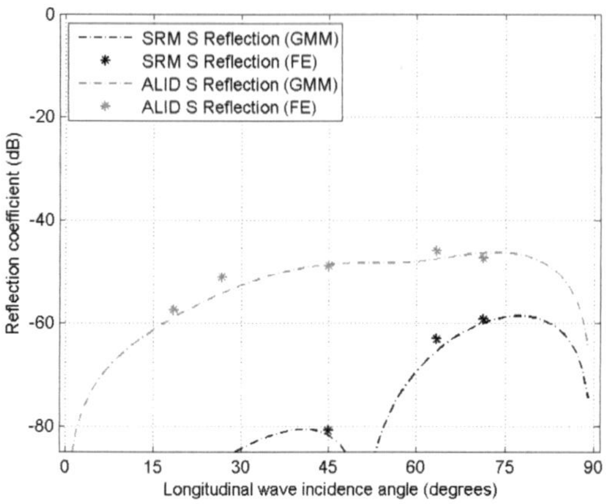

Figure 4. Global Matrix Method (GMM) and numerical (FE) reflection coefficients as a function of incident angle for the reflected shear wave mode (S) from SRM and ALID absorbing boundaries for an incident longitudinal bulk wave.

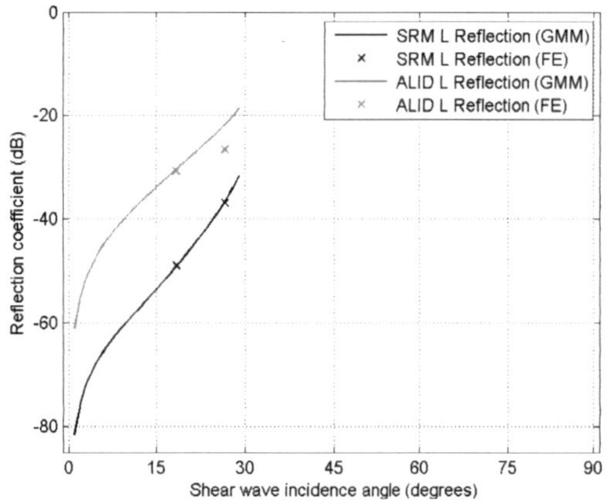

Figure 5. Global Matrix Method (GMM) and numerical (FE) reflection coefficients as a function of incident angle for the reflected compression wave mode (L) from SRM and ALID absorbing boundaries for an incident shear bulk wave.

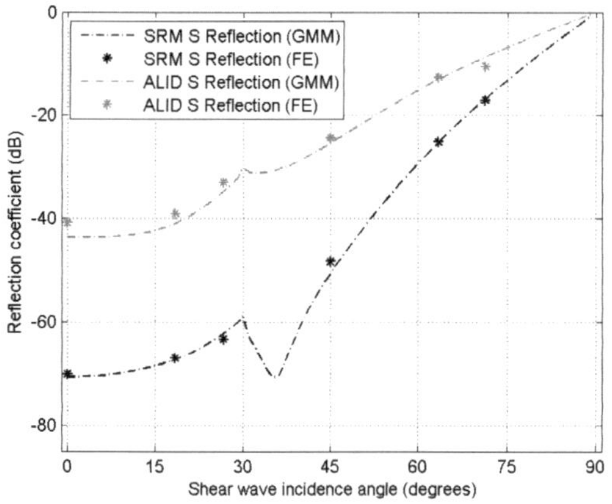

Figure 6. Global Matrix Method (GMM) and numerical (FE) reflection coefficients as a function of incident angle for the reflected shear wave mode (S) from SRM and ALID absorbing boundaries for an incident shear bulk wave.

From these figures it is apparent that the SRM out performs ALID across all angles of incidence due to the attenuation of the terms in the stiffness matrix.

There is excellent agreement between the numerical and analytical solutions for both the ALID and SRM, confirming the reliability of the analyses. Performance at low angles of incidence is significantly better than at higher glancing angles; consistent with other absorbing layer techniques. This confirms that the angle of incidence plays a critical role in the selection of SRM variables. Performance at high angles will always render a less acceptable solution (however it is often the case that these angles are not of concern in simulations). The presence of critical angles generates evanescent waves propagating parallel to the back boundary of the SRM. The angle at which this occurs is not necessarily obvious since it is due to a net effect from all the layers combined. This is illustrated in figure 5 whereby after the critical angle for an incident shear wave, the reflected longitudinal component is not radiated back into the AoS.

4. Optimising SRM performance

When building any absorbing region, there are a number of variables at the disposal of the user. Generic rules can be applied to absorbing region techniques that will ensure good performance. Despite the success of the absorbing boundary methods, it is noticed that for three-dimensional models, the additional contribution of the absorbing layers to the spatial domain is relatively large. The SRM helps to overcome this issue by significantly reducing the size of the spatial domain, however, through the use of optimisation functions, highly specialised absorbing regions can be designed that will result in further improvements.

This section will discuss methods to optimise the SRM absorbing layer. To do this, the requirements of the layer must first be defined. Here, the requirement is that the SRM absorbing layer will be optimised to absorb incident waves at a particular angle of incidence. This must be achieved within the smallest thickness of absorbing material possible.

Optimisation functions can be used to find the optimum input parameters that meet a given condition, for a given function. Here, the optimisation function 'fminsearch' provided by MATLAB is used, [15]. The function looks to calculate the input parameters that will achieve the smallest possible reflection coefficient at a given angle of incidence. This is achieved by varying all input values until the minimum reflection coefficient at the specified angle has been found.

If an absorbing layer is being built for a specific scenario, it may be necessary to put restrictions in place. This is done by fixing one or more of the variables, for example, ω, N and L, which will be constants when using an explicit FE model.

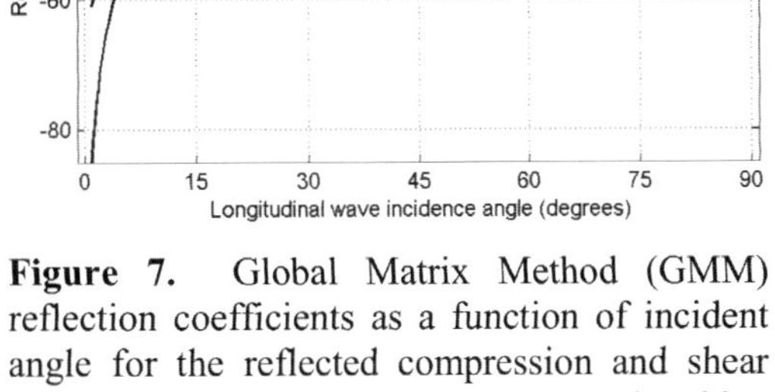

Figure 7. Global Matrix Method (GMM) reflection coefficients as a function of incident angle for the reflected compression and shear wave modes (L and S) from a SRM absorbing boundary that is optimised for an incident longitudinal bulk wave at 0°.

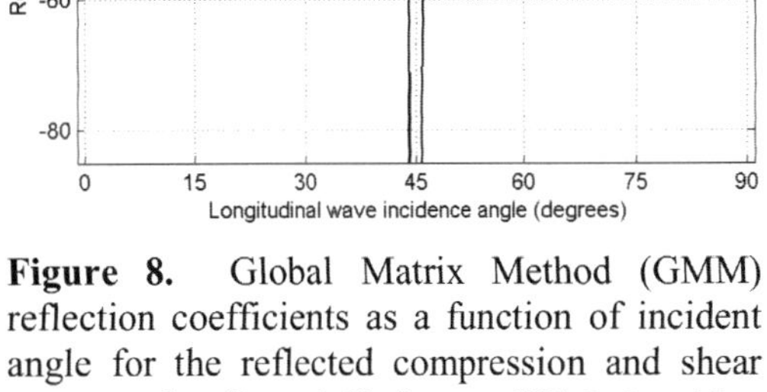

Figure 8. Global Matrix Method (GMM) reflection coefficients as a function of incident angle for the reflected compression and shear wave modes (L and S) from a SRM absorbing boundary that is optimised for an incident longitudinal bulk wave at 45°.

Figure 7 and figure 8 show the reflection coefficients from an SRM absorbing region for a model which has longitudinal and shear wavelengths of 1 and 2 respectively. The SRM thickness is fixed to a length of $\lambda_{inc}/4$ and the optimisation function has been used to calculate the input parameters such that a minimal reflection is achieved for two specific angles of incidence, 0° and 45° respectively.

From figure 7 and figure 8 it can be seen that the optimisation function can be used to build highly specialised absorbing boundaries that operate within significantly reduced spatial domains. This approach provides a means to significantly reduce the computational burden associated with three-dimensional models, by tailoring the high level of performance to within a narrow angular range.

To use the optimisation function correctly, it is important to avoid scenarios where local minima are found over the global minima. To achieve this, the optimisation function requires starting values for the inputs, before finding the optimal input values themselves. These should be equal to those given for the generic case described in section 2.2. This will help to avoid finding the local minima over the global values.

5. Conclusions

The SRM is a new absorbing boundary method that has been developed to operate within a significantly reduced spatial domain. Furthermore, the SRM can be easily implemented into commercial FE codes and does not require specific programming or functionality. All absorbing boundary variables can be controlled from the stiffness and damping matrices.

The reduction in stiffness is defined as a function of the known physical properties of the model, maximum damping values are a function of the incident frequency and the thickness of the absorbing region is of the order of $1.5\lambda_{inc}$.

A two dimensional analytical model based upon the Global Matrix method was shown to have excellent agreement with numerical simulations. Through use of an optimisation function, a highly specialised SRM can be developed for application towards specific scenarios. The optimisation function looks to reduce the magnitude of unwanted reflections that occur at a particular angle of incidence. This allows for a high performing absorbing boundary that can operate within a spatial domain that is further reduced; however, this is only applicable for the designed scenario.

By applying the SRM, computational resource can be focused on a highly accurate representation AoS. This is practically advantageous when considering three-dimensional problems where even a significantly reduced spatial domain can consist of tens of millions of degrees of freedom to be solved.

Acknowledgments

The author would like to acknowledge the help provided by Dr Wonjae Choi, Dr Elizabeth A. Skelton, Prof. Richard V. Craster at Imperial College London, the NDE Research team at Rolls-Royce, the Research Centre in Non-Destructive Evaluation (RCNDE), and the Engineering and Physical Sciences Research Council (EPSRC) for their financial support during the project.

References

[1] Smith W D 1975 *Geophys. J. R. Astron. Soc.* **42** 747-68
[2] Ludwig R and Lord W 1988 *IEEE Trans. Ultrason. Ferroelectr. Freq. Contr.* **35** 809-20
[3] Rajagopal P, Skelton E A, Choi W, Lowe M J S and Craster R V 2012 *IEEE Trans. Ultrason. Ferroelectr. Freq. Contr.* **59** 1239-52
[4] Mahaut S, Leymarie N, Poidevin C, Fouquet T and Dupond O 2011 *Insight* **53** 664-7
[5] Wilcox P D and Velichko A 2010 *J. Acoust. Soc. Am.* **127** 155-65
[6] Bettess P 1992 *Infinite Elements* (Penshaw Press)
[7] Givoli D and Keller J B 1990 *Wave Motion* **12** 261-79
[8] Israeli M and Orszag S A 1981 *J. Comp. Phy.* **41** 115-35
[9] Berenger J P 1994 *J. Comp. Phys.* **114** 185-200
[10] Semblat J F, Lenti L, and Gandomzadeh A 2011 *Int. J. for Num. Methods in Eng.* **85** 1543-63
[11] Rajagopal P, Drozdz M, Skelton E A, Lowe M J S, and Craster R V 2012 *NDT&E Int.* **51** 30-40
[12] Pettit J R and Walker A and Cawley P and Lowe M J S 2014 *Ultrasonics* **54** 1868-79
[13] Lowe M J S 1995 *IEEE Trans. Ultrason. Ferroelectr. Freq. Contr.* **42** 525-42
[14] ABAQUS v 6.11 2011 *Analysis User's Manual* 2011 http://www.simulia.com/
[15] MATLAB R2007a 2012 *The MathWorks Inc.* United Kingdom

Behaviour of ultrasonic waves in porous rigid materials: an anisotropic Biot-Attenborough model

H Aygün and C Barlow

Maritime and Technology Faculty, Southampton Solent University, Southampton, SO14 0RD, UK.

E-mail: haydar.aygun@solent.ac.uk

Abstract. The anisotropic pore structure and elasticity of cancellous bone cause wave speeds and attenuation in cancellous bone to vary with angle. Anisotropy has been introduced into Biot theory by using an empirical expression for the angle-and porosity-dependence of tortuosity. Predictions of a modified anisotropic Biot–Attenborough theory are compared with measurements of pulses centred on 100 kHz and 1 MHz transmitted through water-saturated porous samples. The samples are 13 times larger than the original bone samples. Despite the expected effects of scattering, which is neglected in the theory, at 100 kHz the predicted and measured transmitted waveforms are similar.

1. Introduction

Understanding the propagation of acoustic waves through cancellous bone is an important pre-requisite to improving the diagnosis of osteoporosis by ultrasound. Bone essentially has two types of structure, both having the same mineralised collagen composition. Cortical bone has porosity less than 30% and may generally be considered to be solid; cancellous bone has porosity greater than 30% and consists of a complex open-celled porous network of rod- and plate-shaped elements termed trabeculae. The porosity of human cancellous bone ranges between 70% and 95%, the remaining volume being perfused with bone marrow. In the adult human vertebral body for example, both horizontal and vertical trabeculae range from 50–120 μm in thickness, and spaced at intervals of between 1200 – 5000 μm and 700 and 2000 μm respectively [1]. During childhood, more bone is added than is being taken away.

During early adulthood, the amounts removed and added are the same. If however, more bone is removed than is being added, we have a condition called **osteoporosis** which literally means ‘porous bone’ and describes a period of largely asymptomatic bone loss leading to skeletal fragility and increased risk of fracture. It is caused by hormonal imbalance (oestrogen & testosterone) and long-term cortico-steroid use. It is also caused by low bone mass, as well as a weakened structure [2] and [3]. One in three women and one in five men over the age of 50 will break a bone attributed to osteoporosis. It is second only to cardiovascular disease as a global healthcare problem [4]. Osteoporosis leads to nearly 9 million fractures annually worldwide [5], and over 300,000 patients present with fragility fractures to hospitals in the UK each year [6]. Direct medical costs from fragility fractures to the UK healthcare economy were estimated at £1.8 billion in 2000, with the potential to increase to £2.2 billion by 2025, and with most of these costs relating to hip fracture care [7]. To improve the prediction of fracture risk by ultrasound it is important to understand the propagation of acoustic waves through cancellous bone. The theory that is mostly used for investigation of acoustic wave propagation in cancellous bone is Biot theory which was specifically developed to describe acoustic wave propagation in fluid-saturated porous elastic media [8] and [9]. The Biot theory allows for an arbitrary microstructure, with separate motions considered for the solid elastic framework (bone) and the interspersed fluid (marrow), induced by the ultrasonic wave, and also includes energy loss due to viscous friction between solid (bone) and fluid (marrow). The Biot theory predicts a shear wave. The Biot model has since been used extensively to describe the wave motion in trabecular (cancellous) bone [10], [11], [12], [13], [14], [15] and [16]. Attenborough *et al.* [17] presented tortuosity deduced from audio-frequency measurements in air-filled cancellous bone replicas and showed that there was strong anisotropy.

Laugier and Haiat [18] have presented an extensive view of the mathematical and numerical models to understand Quantitative ultrasound (QUS) potential and the types of variables that can be determined by QUS in order to characterize bone strength. A reasonable protocol has been presented for case-finding purposes, which relies on a combined assessment of clinical risk factors (CR.F) and heel QUS by Hans and Krieg [19]. The clinical applications of QUS in the detection and management of osteoporosis and osteoporosis-related fractures have been reviewed. Foiret *et al.* [20] have used ultrasound waves to characterize and determine the structural and material properties of cortical bone. The frequency spectrum of Lamb type of waves is the input data of the method of parameters identification, based on a least mean square algorithm. Resonant ultrasound spectroscopy (RUS) has been used to measure the anisotropic elasticity of the cortical bone despite the strong viscoelastic damping. RUS does not require a minimum size for the tested specimens, it may be appropriate for the characterization of small bone specimens. Normal and pathological bone material properties can be investigated by using RUS.

Previous work on the influence of anisotropic pore structure and elasticity in cancellous bone has been extended by developing an anisotropic Biot–Allard model allowing for angle dependent tortuosity and elasticity by Aygün *et al.* [21]. The extreme angle dependence of tortuosity corresponding to the parallel plate microstructure used by Hughes *et al.* [15] has been replaced by angle dependent tortuosity values based on data for slow wave transmission through air-filled bone replicas. Audio-frequency data obtained at audio-frequencies in air-filled bone replicas are used to derive an empirical expression for the angle-and porosity-dependence of tortuosity. Most recently, Aygün *et al.* [22, 23] have transmitted ultrasonic signals through water saturated stereolithograpical bone replicas. It is found that the expected occurrence of scattering does not cause significant discrepancies between predictions and data at 100 kHz (which would be equivalent to 1.3 MHz in real bone), perhaps as a consequence of the fact that the samples behave as low pass filters. Scattering should be even more important at 1 MHz (equivalent to 13 MHz in real bone) where the fast and slow wavelengths are 3 mm and 1.5 mm respectively.

Aim of this work is to investigate the dependence of ultrasonic wave propagation upon the material and structural properties of porous rigid materials. Stereolithograpical (see Figure 1) bone replicas and ceramic have been used for the investigation of the influences of perforation and thinning in cancellous bone on the acoustical and mechanical properties of the bone structure.

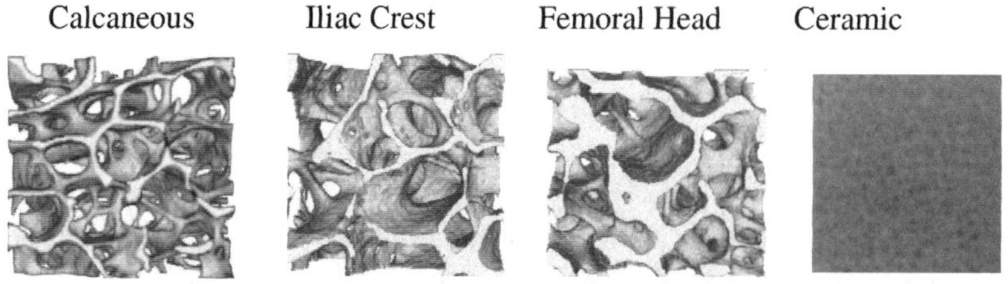

Figure 1. Views of three stereolithographical (STL) bone replicas and a porous ceramic.

2. Theory

A porous rigid sample of length L is subjected to an incident ultrasonic wave in fluid (water), P^i (see Figure 2). Part of incident wave is reflected back into the fluid, P^r, while other part is transmitted through the sample, P^t. Fellah *et al.* have presented an analytical model in order to describe the viscous interaction between fluid and a porous elastic structure. The Fourier transform of the transmitted field is given by Fellah *et al.* as:

$$P_3(x,\omega) = \tilde{T}(\omega)\exp\left(-j\omega\frac{(x-L_{TP})}{c_0}\right)\varphi(\omega), \qquad x \geq L_{TP} \tag{1}$$

where $\varphi(\omega)$ is the Fourier transform of the incident field ($P^i(t)$), $\tilde{T}(\omega)$ is the Fourier transform of the transmission kernel, ω is the angular frequency of motion, c_0 is the speed of sound in fluid, and L_{TP} is the transmission path. A more detailed consideration of the transformed field can be found in the paper by Fellah *et al.* The transmission coefficient $T(\omega)$, which is the Fourier transform of $\tilde{T}$.

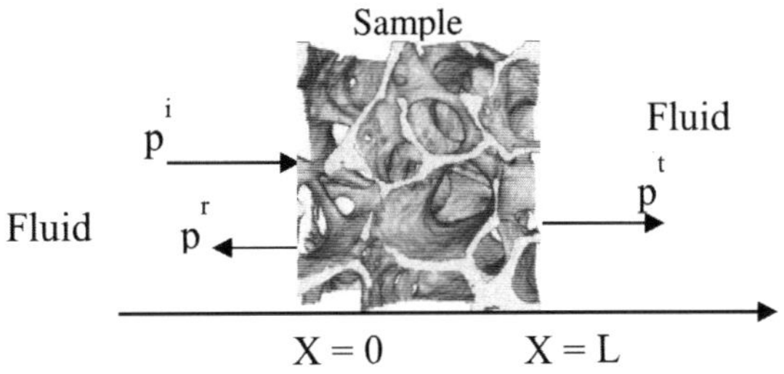

Figure 2. Geometry of a porous rigid material

Aygün *et al.* [21] have introduced a transverse anisotropy into Biot-Allard model by allowing angle-and-porosity dependent tortuosity, and angle-dependent elasticity. Tortuosity s known to have an important influence on high frequency sound propagation in fluid saturated porous media. A heuristic form for porosity- and angle- dependent tortuosity is proposed by Aygün *et al.* [21] *as*:

$$\alpha_\infty = 1 - r\left(1 - \frac{1}{\phi}\right) + k\cos^2(\theta) \tag{2}$$

where ϕ is the porosity, θ is the variable between 0° and 90°, r and k can be considered adjustable. A range of possible values of r and k have been found by comparing predictions of equation (2) for θ = 0° and 90° respectively with values deduced from air-filled replica [17] of known porosity. Values of r and k are found by solving the resulting simultaneous equations.

To predict transmission through an anisotropic poroelastic sample it is necessary to allow for elastic anisotropy also. The dependence of skeletal frame modulus (Young's modulus, E_b, Bulk Modulus, K_b, and rigidity modulus, μ_b) in terms of bone volume fraction $(1 - \phi)$ and the Young's modulus of the solid material of the frame (E_s) are given by Williams [24]:

$$E_b = E_s(1-\phi)^n,$$
$$K_b = E_b/(1-2\upsilon_b),$$
$$\mu_b = E_b/(1+2\upsilon_b), \tag{3. a, b, c}$$

where v_b is the Poisson's ratio of frame, and the exponent n varies from 1 to 3 according to Gibson [25], depending on the angle (θ) with respect to the dominant structural orientation according to $n = n_1$

$\sin^2(\theta) + n_2 \cos^2(\theta)$. Values of $n_1 = 1.23$ and $n_2 = 2.35$ are chosen by Lee *et al.*[26] to be consistent with the work of Williams [24].

3. Measurements

The experimental procedure used by Fellah *et al.* [13] has been followed to perform measurements in a water tank (see Figure 1). Two 100 kHz and two 1 MHz plane piezoelectric transducers have been used for experiments. 400 V pulses are provided by a 5058PR Panametrics pulser/receiver. Electronic interference is removed by 1000 acquisition averages. Once we start revolving the bone replica sample around itself, its thickness that ultrasonic wave transmitted through became equal to $L.\cos(\theta)$ where θ is the angle of propagation.

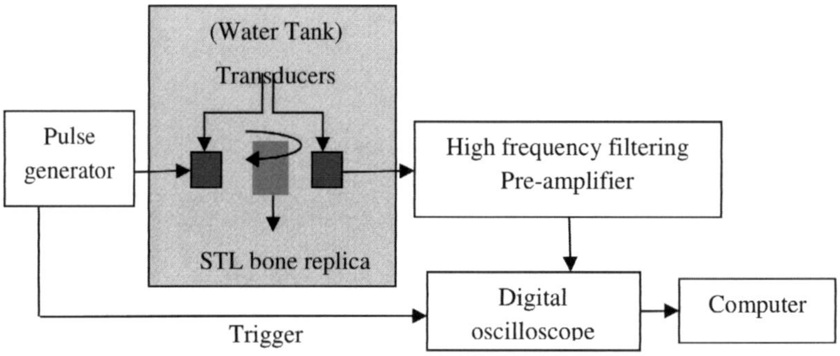

Figure 3: Experimental setup for ultrasonic measurements.

When a wave impinges on a STL bone replica, part of the wave is reflected back. The part of the wave penetrating into the sample undergoes mode conversion into fast and slow components which are transmitted through the STL bone replica. The measurements have been made parallel to the trabeculae direction. The stereolithographical bone replicas used in the measurements are in the form of 57 mm cubes. The incident (reference) signals generated by 100 kHz and 1 MHz transducers and transmitted through fluid (water) are shown in Figure 4a and Figure 5a, and their spectra are shown in Figure 4b and Figure 5b, respectively. The waveform at 100 kHz looks like a complex waveform consisting of more than one signal while the waveform at 1 MHz looks like a sine wave with its amplitude reducing by time.

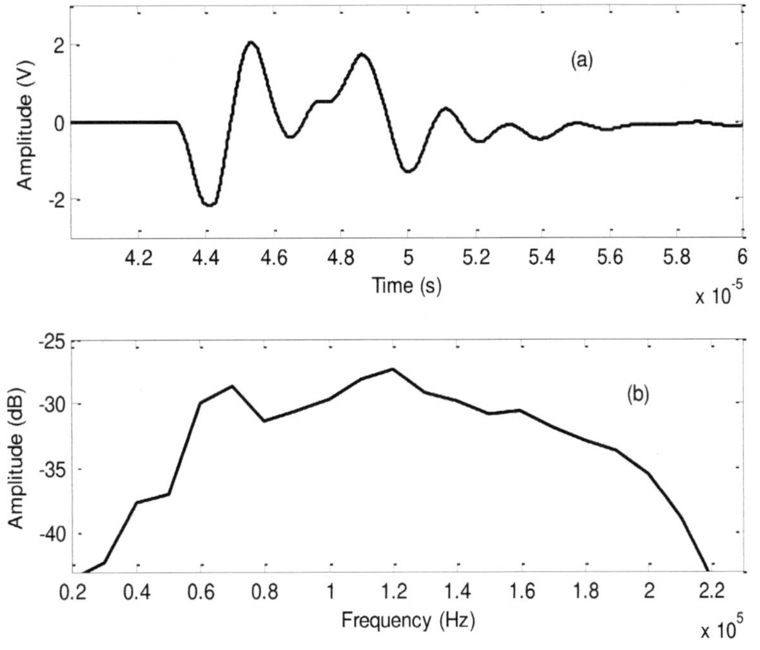

Figure 4. a-) Inicident signal versus time, b-) its spectrum versus frequency at 100 kHz.

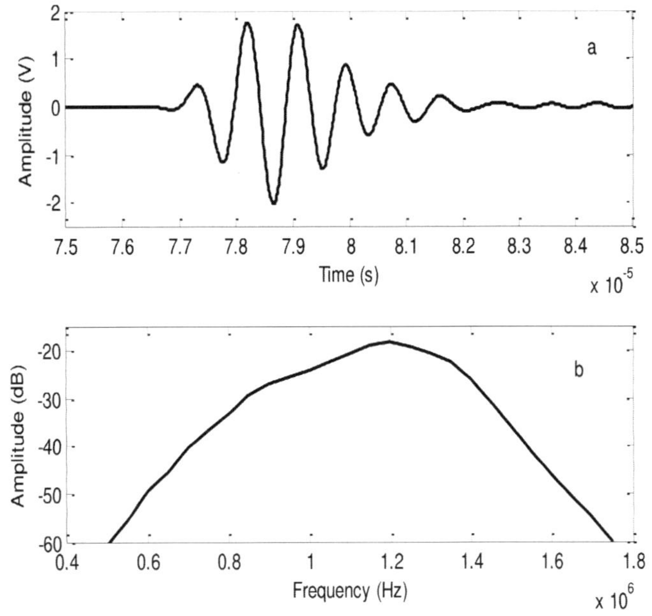

Figure 5. a-) Incident signal versus time, and b-) Its spectrum versus frequency at 1 MHz.

4. Comparisons between predictions and data

Wave transmission through replicas has been predicted at normal angles and results have been compared with measurements performed in a water filled tank. The parameters used in the predictions are listed in Table 1. The elastic moduli of the bone replicas made of resin have been taken to be equal to the elastic modulus of resin which is 6.04 GPa (DSM Somos) and is smaller than the elastic modulus of real bone which is 20 GPa [24] Assuming that the permeability of the bone is 5×10^{-9} m^3 [10], the permeability of bone replicas has been taken to be 13^2 times higher because of the magnification of the actual size of the bone microstructure by 13 times in each direction. The assumed characteristics of the saturating fluid (water) are: density $\rho_f = 1000$ kg/m^{-3}, viscosity $\eta = 10^{-3}$ kg ms^{-1}, speed of sound in water $c_0 = 1490$ m/s. Measured and predicted transmitted signals are travelling through the bone replicas in the same direction as the trabecular alignment (x direction). The incident signal shown in Figure 4a has been used for predictions. Predicted transmitted waves of STL Iliac Crest (ICF), and Femoral Head (FRA) replicas versus time are compared with measured transmitted waves in water at 100 kHz in Figures 6, and 7, respectively. The agreement between predicted and measured transmitted waveforms is reasonable. The initial parts of the measured and predicted transmitted waveforms can be identified as the fast wave arrival while the second and major parts of the transmitted waveforms can be identified as the slow wave contribution. Arrival times for measured and predicted fast waves of ICF bone replica at 100 kHz are 3.4 x10^{-5} s and 3.8 x10^{-5} s respectively. Arrival times for measured and predicted fast waves of FRA bone replica at 100 kHz are 3x10^{-5} s and 3.5x10^{-5} s respectively. A possible reason of late arrival of fast wave might be signal leakage caused by refraction from the CAB replica at higher angles and transmission paths that fall outside the receiving transducers. Fifteen times of enlargement of the original sample in 3 directions might be another reason for late arrival. It is not possible to determine the exact arrival time for slow waves because of overlapping between fast and slow waves.

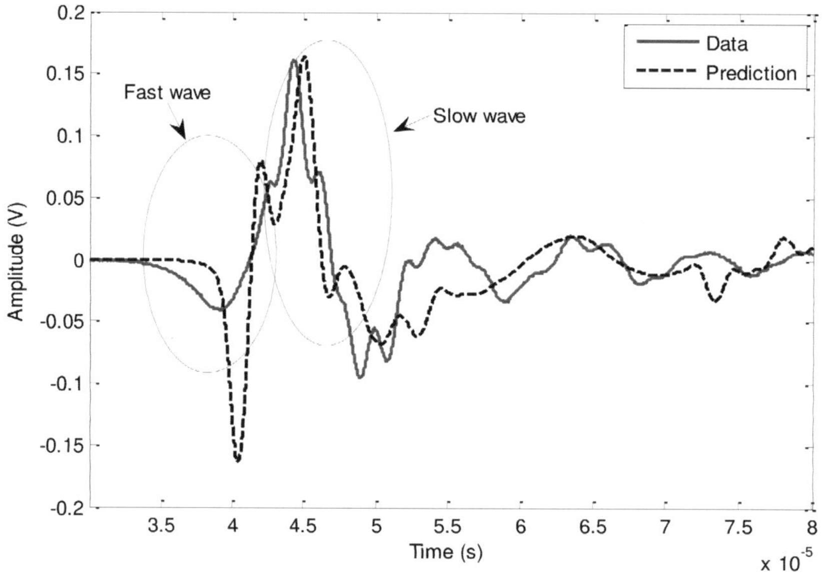

Figure 6. Comparison of predicted and measured transmitted waveforms through a water-saturated ICF bone replica at 100 kHz at normal incidence.

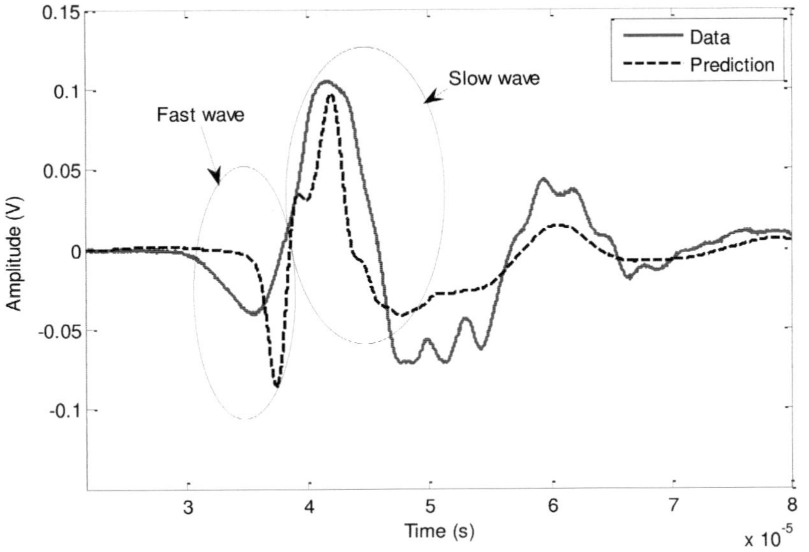

Figure 7. Comparison of predicted and measured transmitted waveforms through a water-saturated FRA bone replica at 100 kHz at normal incidence.

Transmitted waveforms at the different oblique angles versus time have been measured in order to understand the dependence of the wave transmission on the propagation angle. Measured transmitted waveforms of CAB bone replica subjected ultrasonic propagation at different angle of propagation are quite different from each other as seen in Figure 8. Generally, measured and predicted transmitted waveforms through CAB bone replica are very similar. The arrival time of the ultrasonic waves changes when the angle of propagation was varied. Since the tortuosity equation used in this paper depends on the angle of propagation, varying the angle changes the tortuosity. The arrival time of the ultrasonic wave depends on the quantity of the bone in the path of the wave transmission. Only two parameters, Poisson's ratio of frame and viscous characteristic length, were adjusted in order to obtain the 'best-fit'. In particular, the predictions are very sensitive to the assumed values of viscous characteristic length. The 'best-fit' characteristic length values (Table 1) for the replicas are about 13 times those found for real bone in the literature i.e. between 5 and 10 μm [13, 14].

Table 1: Input parameters for Stereolithograpical (STL) bone replicas.

Parameters	Iliac Crest, ICF	Femoral Head, FRA	Calcaneus, CAB	Porous Ceramic
Density, ρ_s, kg/m^3	1233.4	1227	1171	950
Young's modulus, E_s, GPa	6.04	6.04	6.04	2.6
Poisson's ratio of solid, ν_s	0.30	0.30	0.30	0.29
Poisson's ratio of frame, ν_b	0.36	0.40	0.34	0.35
Porosity, ϕ	0.8386	0.7426	0.8822	0.62
Permeability, k_0, m^3	845×10^{-9}	845×10^{-9}	845×10^{-9}	5×10^{-10}
Form factor, c	1.2	1.2	1.2	1.2
Form factor, c'	c/2	c/2	c/2	c/2
Viscous characteristic length, L, μm	100	60	150	38
r	0.888	0.591	0.816	0.56
k	0.468	0.684	0.574	0.44

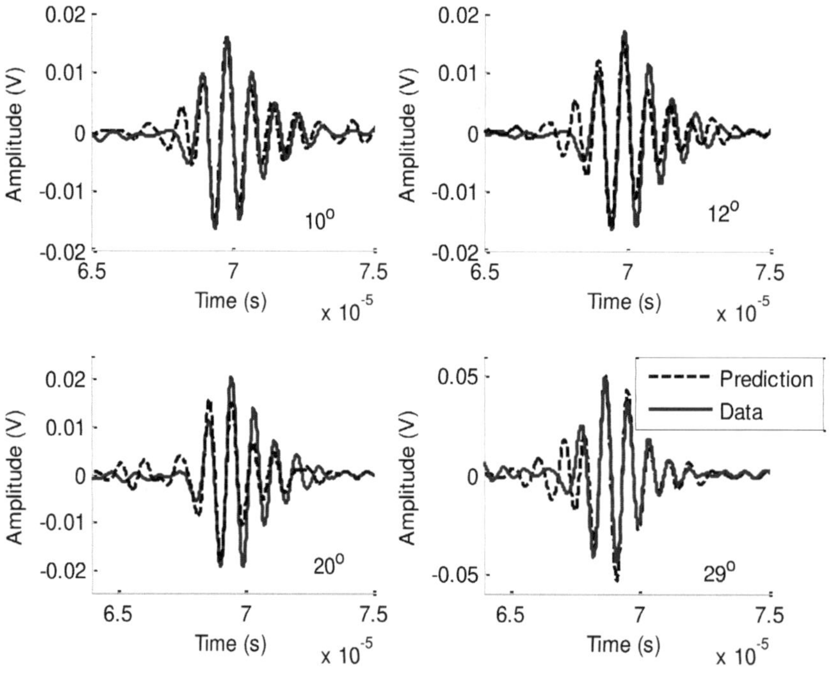

Figure 8. Measured and predicted transmitted waveforms through CAB bone replica versus time with 1MHz transducers at 10°, 12°, 20°, and 29°.

Journal of Physics: Conference Series **581** (2015) 012006 doi:10.1088/1742-6596/581/1/012006

Predicted transmission coefficient as a function of frequency for ceramic is shown in Figure 9. Porous ceramic is obtained by mixing clay and plastic then burning the plastic in a kiln at Laboratory of Acoustics and Thermal Physics at K. U. Leuven. Porous ceramic used for measurements is in the form of 65 mm squares with 30 mm thickness. Predicted and measured transmitted waveform in the porous ceramic at oblique incidence (25°) is presented as a function of propagation time in Figure 9. There is reasonable agreement between predictions and data. It seems that less than 3 % of the incident wave amplitude is transmitted through porous ceramic at 1 MHz. Most of the ultrasonic wave is reflected back into the water. Generally measured and predicted transmitted waveforms are similar except for the initial parts of the transmitted waveforms.

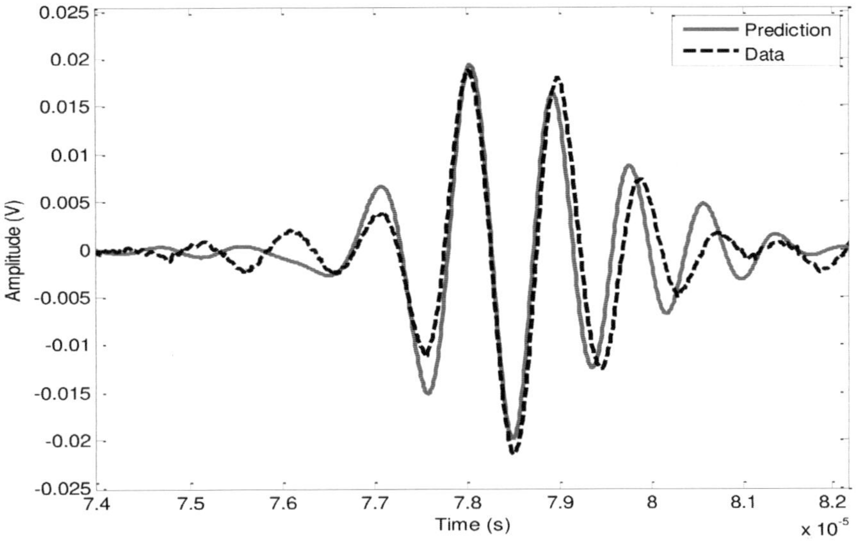

Figure 9. Measured and predicted transmitted waveforms through porous ceramic versus time at 25°.

5. Conclusion and further work

Predictions of a modified anisotropic Biot-Allard theory which neglecting scattering, have been compared with measurements of pulses centered on 100 kHz and 1 MHz transmitted through water saturated porous rigid samples at normal and oblique angles. The predictions and data are in reasonable agreement.

Further works need to be done on porous solid materials (ceramics) with different porosity values to investigate the Osteoporosis by using structural borne vibration. A sinusoidal vibration should be applied to porous samples. This will cause a structural borne sound wave to propagate through the samples. The resulting vibration should be measured using either an ultrasound probe or an accelerometer. Based on these results, a hand held, low cost and portable device should be designed and developed to detect the Osteoporosis in the bone with the aim of improving early detection rates.

The chief debilitating consequence of osteoporosis is fracture, and early detection of the condition can allow interventions reducing the likelihood of occurrence [25]. As hip fracture invariably leads to hospital admission, early detection of the condition could significantly reduce the overall long term treatment costs by reducing admission to hospital [26]. Due to anticipated low costs of the product, it is anticipated that access to early screening would improve with increased overall take up by clinics compared to other technologies.

Acknowledgement

This work has been supported by Leverhulme Grant: F/00 181/N which provided for collaboration with the Laboratory of Acoustics and Thermal Physics at Leuven where the data reported here were obtained.

References

[1] Thomsen J S, Ebbesen E N, Mosekilde L 2002, Age-related differences between thinning of horizontal and vertical trabeculae in human lumbar bone as assessed by a new computerized method. Bone **31**, 136-142.

[2] Prentice A 1997 Is nutrition important in osteoporosis? *Proceedings of the Nutrition Society*, **56**: 357-367.

[3] Cooper C, Campion G, and Melton L J, 1992 Hip fractures in the elderly: a world-wide projection. *Osteoporos Int* **2**:285.

[4] World Health Organization 2004, Scientific group on the assessment of osteoporosis at primary health care level, Summary Meeting Report, Brussels. Javadi Y, Pirzaman H S, Raeis M H, Najafabadi M A 2013 *Materials and Design* **49** 591-601

[5] Johnell O, and Kanis J A, 2006 An estimate of the worldwide prevalence and disability associated with osteoporotic fractures. Osteoporosis International **17**: 1726–33.

[6] British Orthopaedic Association, 2007 The care of patients with fragility fracture.

[7] Burge R T, Worley D, and Johansen A, 2001 The cost of osteoporotic fractures in the UK: projections for 2000–2020. *J. Medical Economics* **4**: 51–52.

[8] Biot M A 1956 Theory of propagation of elastic waves in a fluid saturated porous solid, I. Low frequency range. *J. Acoust. Soc. Am.* **28** 168–1178.

[9] Biot M A 1956 Theory of propagation of elastic waves in a fluid saturated porous solid, II. High frequency range. *J. Acoust. Soc. Am.* **28** 179–191.

[10] McKelvie M L and Palmer S B 1991 The interaction of ultrasound with cancellous bone. *Phys. Med. Biol.* **36** 1331-1340.

[11] Hosokawa A, and Otani T 1997 Ultrasonic wave propagation in bovine cancellous bone. *J. Acoust. Soc. Am.***101** 558-562.

[12] Haire T J, and Langton C M, 1999 Biot theory: A review of its application on ultrasound propagation through cancellous bone. *Bone* **24** 291–295.

[13] Fellah Z E A, Chapelon J Y, Berger S, Lauriks W, and Depollier C 2004 Ultrasonic wave propagation in human cancellous bone: Application of Biot theory. *J. Acoust. Soc. Am.* **116** 61–73.

[14] Sebaa N, Fellah Z, Fellah M, Ogam E, Wirgin A, Mitri F, Depollier C, and Lauriks W 2006 Ultrasonic characterisation of human cancellous bone using the Biot theory: Inverse problem. *J. Acoust. Soc. Am.* **120** 1816-1824.

[15] Hughes E R, Leighton T G, White P R and Petley G W 2007 Investigation of an anisotropic tortuosity in a Biot model of ultrasonic propagation in cancellous bone. *J. Acoust. Soc. Am.* **121** 568–574.

[16] Pakula M, Padilla F, Laugier P and Kaczmarek M 2008 Application of Biot's theory to ultrasonic characterization of human cancellous bones: determination of structural, material, and mechanical properties. *J. Acoustic Soc Am.* **123** (4): 2415-2423.

[17] Attenborough K, Qin Q, Fagan M J, Shin H-C, and Langton C M 2005 Measurements of tortuosity in stereolithographical bone replicas using audio-frequency pulses. *J. Acous. Soc Am.* **118** 2779-2782.

[18] Laugier P and Haiat G, 2011. Bone quantitative ultrasound. Springer.

[19] Hans D and Krieg M A 2008 The clinical use of quantitative ultrasound (QUS) in the detection and management of osteoporosis. IEEE **55** (7), 1529-1538.

[20] Foiret J, Minonzio J G, Talmant M, and Laugier P 2013 Determination of bone properties from Lamb type of waves. *J. Acoust. Soc. Am.* **133** (5), 3585-3592.

[21] Aygün H, Attenborough K, Postema M, Lauriks W, and Langton C M, 2009, Predictions of angle dependent tortuosity and elasticity effects on sound propagation in cancellous bone. *J. Acoust. Soc. Am.* **126** (6), 3286-3290.

[22] Aygün H., Attenborough K., Lauriks W., and Langton C. M., 2010, "Ultrasonic wave propagation in Stereo-lithographical bone replicas." *J. Acoust. Soc. Am.* **127** (6), 3781 – 3789.

[23] Aygün H., Attenborough K., Lauriks W., Rubini P.A., and Langton C. M., 2011 Wave propagation in stereo-lithographical bone replicas at oblique incidence. *Applied Acoustics*. **72** (7) 458–463.

[24] Williams, J. L. 1992 "Ultrasonic wave propagation in cancellous and cortical bone: predictions of some experimental results by Biot's Theory," J. Acoust. Soc. Am. **92**, 1106–1112.

[25] Gibson, L.J. (1985) The Mechanical Behaviour of Cancellous Bone. *Journal of Biomechanics*, **18**, 317-328.

[26] Lee K I, Roh H-S, and Yoon S W 2003 Acoustic wave propagation in bovine cancellous bone: Application of the Modified Biot–Attenborough model. *J. Acoust. Soc. Am.* **114** 2284-2293.

[27] Tucci JR. 2006, "Importance of early diagnosis and treatment of osteoporosis to prevent fractures." *Am J Manag Care.* May; **12**(7 Supply):S181-90.

[28] King A B, Saag K G, Burge R T, Pisu M, Goel N, 2005, "Fracture Reduction Affects Medicare Economics (FRAME): impact of increased osteoporosis diagnosis and treatment." *Osteoporos Int*. Dec; **16** (12):1545-57.

Journal of Physics: Conference Series **581** (2015) 012007 doi:10.1088/1742-6596/581/1/012007

Quantitative imaging of acoustic reflection and interference

Robert Malkin[1], Thomas Todd[2] and Daniel Robert[1]

[1]School of Biological Sciences, University of Bristol, England, BS8 1TQ
[2]Faculty of Engineering, University of Bristol, England, BS8 1TR

E-mail: r.e.malkin@gmail.com

Abstract. This paper presents a method for time resolved quantitative imaging of acoustic waves. We present the theoretical background, the experimental method and the comparison between experimental and numerical reconstructions of acoustic reflection and interference. Laser Doppler vibrometry is used to detect the modulation of the propagation velocity of light, c, due to pressure-dependant changes in the refractive index of air. Variation in c is known to be proportional to variation in acoustic pressure and thus can be used to quantify sound pressure fluctuations. The method requires the laser beam to travel through the sound field, in effect integrating pressure along a transect line. We investigate the applicability of the method, in particular the effect of the geometry of the sound radiator on line integration. Both experimental and finite element reconstructions of the sound field are in good agreement, corroborating punctual pressure measurements from a precision microphone. Spatial limitations and accuracy of the method are presented and discussed.

1. Introduction
Quantified reconstruction of acoustic fields is commonly done using microphone arrays. These arrays tend to be expensive, bulky and uncertainty arises at higher frequencies where microphones can interfere with the sound field. Schlieren photography is a well-known method of visualising acoustic fields but lacks the ability of simple pressure field quantification [1]. The calculation of instantaneous sound pressure using light intensity measurements has been shown to be sensitive and accurate [2]. The use of laser Doppler vibrometry to measure acoustic fields has received prior attention [3]–[8], including tomographic 3D field reconstruction [9]. Previous research has shown that quantified sound field visualisation is possible but the implementation has been complex and results difficult to interpret. Here, we present a simple method (accessible to the non-specialist) of calculating the instantaneous pressure in air and show the importance of using an appropriate sound source for quantified field reconstruction.

2. Theory of refracto-vibrometry
The refractive index, n, of a medium is defined as;

$$n \equiv \frac{c_0}{v} \quad (1)$$

With c_0: velocity of light in a vacuum and v: the velocity of light in the medium. The refractive index of air is influenced by pressure, temperature, humidity and frequency (although the dispersive effect is relatively minor in standard conditions) [10]–[12]. In a simple adiabatic system a change in pressure results in a corresponding change in n. The resulting changes in propagation velocity, v, can be measured using a laser Doppler vibrometer (LDV) [4], [5], [13]. The technique of refracto-vibrometry has primarily been used for acoustic pressure visualisation in air and water [9], [13]–[16]. The principle of the measurement technique is described by Nakamura *et al* [17] and is illustrated in Figure 1.

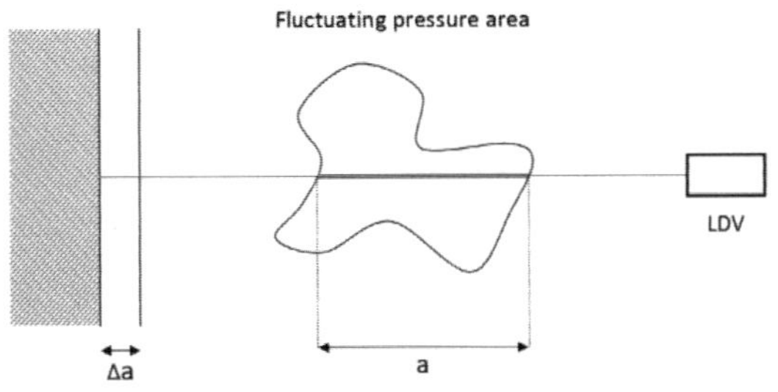

Figure 1. Using LDV, a rigid wall (shaded structure on the left) appears to be displaced by a distance of Δa . This is due to the modulation of the velocity of light propagation within the length a within the fluctuating pressure area.

Monochromatic laser light travelling through a region of sound pressure of length a, results in a corresponding variation in the magnitude of n. The laser light exits the region and, if reflected from a rigid wall, travels back through the region and back the LDV. This modulated path appears identical to a physical displacement of the reflective rigid wall, Δa.

$$n \cdot \Delta a = \Delta n \cdot a \tag{2}$$

Where the apparent displacement of the rigid wall, Δa, is given by integrating the LDV velocity (with respect to time) output signal. It has been shown that the pressure-refraction relation may be expressed in terms of temperature, pressure and relative humidity [18]. It is formalised by the expression:

$$n = 1 + \frac{7.86 \cdot 10^{-4} \cdot P}{273 + T} - (1.50 \cdot 10^{-11}) \cdot RH(T^2 + 160) \tag{3}$$

With P: pressure in kPa, T: temperature in °C and RH: relative humidity in %. At 1 atmosphere of pressure, 20°C temperature and a relative humidity of 50%, equation (3) gives n a value of 1.0002714. This relation between pressure and n allows us to relate the (observed) apparent displacement of the rigid wall to the pressure in the region of interest. Assuming an adiabatic process,

$$\frac{\Delta n}{\Delta P} = \beta = 2.68 \cdot 10^{-9}\ per\ Pa \tag{4}$$

Combining equations (2) and (4);

$$\Delta P = \frac{\Delta n}{2.68 \cdot 10^{-9}} = \frac{n}{a} \cdot \frac{\Delta a}{\beta} \tag{5}$$

The method of operation requires the laser to pass through the length of the region, *a*, meaning that the resulting velocity change is the integration of the average pressure distribution along the length of the path. As the laser travels from the LDV to the reflector and back, the resulting measured velocity from the LDV is an integration along the whole path length. The pressure field is likely to be variable along this length, therefore quantification of the field using equation (5) is complicated as the length *a* is unknown, as detailed by Olsson & Tatar in 2006 [7]. These authors noted that "any quantitative data obtained when measuring 3D sound fields using laser vibrometry must be viewed with some scepticism" due to the line integration problem when using a point source. In 2007, however, Olsson showed that the method could be used to measure the pressure magnitude correctly by using a correction factor [19]. While the results were accurate, Olsson's method did not lend itself to simple measurement of sound fields, requiring considerable technical expertise. Zipser *et al* took the approach of using a solid glass partition as a waveguide thereby constraining the region in which the sound was propagating [14]. This approach results in accurate quantification, yet precludes the use of free field conditions.

A rigorous treatment on the effect of pressure modulation on the optical laser path is detailed by, Torras-Rosell *et al* in [9].

3. Theoretical analysis of line integration of acoustic pressure

Here we use a 2D transient finite element analysis (FEA) (Comsol 4.3) model of acoustic pressure at 5, 10 & 20kHz for two acoustic radiators; a point source and a planar loudspeaker (with a width, w, of 150mm, equal to that of the speaker used in experiments), shown in Figure 2.

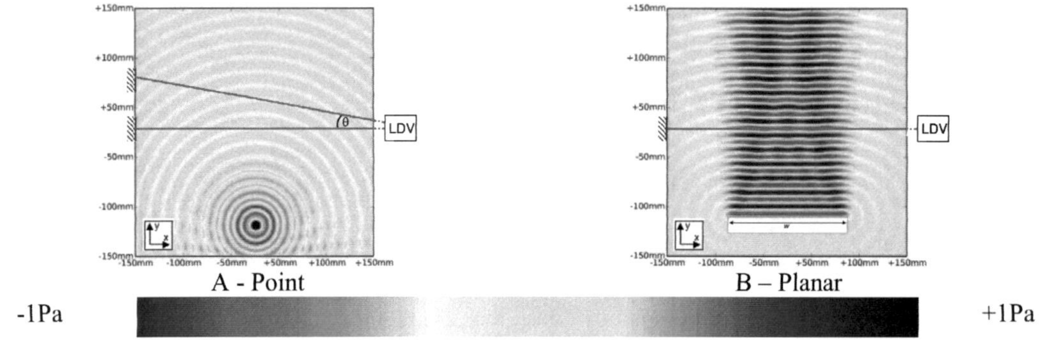

Figure 2. Simulated sound fields at 20kHz showing the position of the LDV, the rigid reflector (hatched area on Y-axis) and the scanning angle θ. A. Visualisation of acoustic field radiating from point source with 2 scan lines drawn. The laser beams from the LDV are shown with deviation angle θ, highlighting the alteration of the laser light path across the sound field. B. Visualisation of acoustic field radiating from a planar source.

In the experimental situation, the laser scanner (Polytec PSV-400) controls the position of the laser beam using rotating mirrors. This results in scan lines that are non-parallel, fanning out from the LDV with angle θ (shown in Figure 2A). The effect of the radiating geometry of the laser beam was investigated by Solodov et al [20]. For the present experimental work, the distance from the LDV to the reflecting wall (2m) was such that the maximum scanning angle was 2.6°. To investigate the effect of these non-parallel scan lines for the planar source case, the average pressure along a scan line (red line, Figure 2B) is compared to a point measurements along the y-axis for y>0mm at x=0mm. The results of this theoretical model-based analysis at 5, 10 and 20kHz are shown in Figure 3 for scanning angles up to 4° at one single phase of the sound pressure. This analysis reveals that for the outermost scan lines, the pressure value may be underestimated by as much as 15% for 20kHz. The error increases with frequency as the wavelength shortens and scan lines 'traverse' an increasing number of wave fronts. This error can be reduced by: 1. increasing the distance between LDV and reflecting wall, 2. reducing the width of the planar loudspeaker *w* and 3. increasing the wavelength. These three optimisations can be used individually or in combination. For the FEA we assume the scan lines as to be parallel. It should be noted that for the non-parallel situation and higher frequencies, wavelengths can become sufficiently short as to cause considerable measurement error.

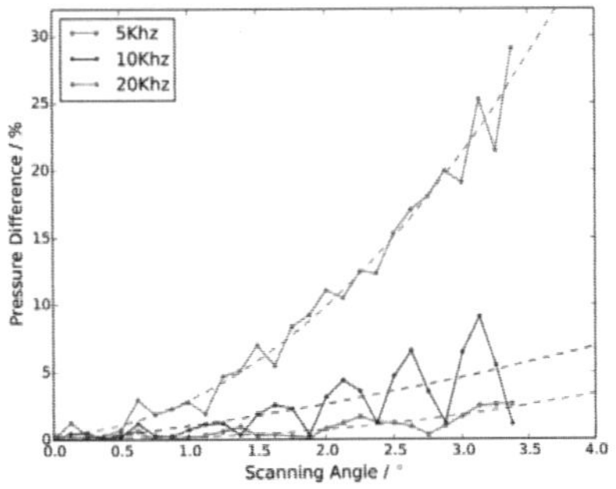

Figure 3. Experimental data of the effect of non-parallel scan lines on actual and line averaged pressure values taken at a single phase of sound pressure. Dashed lines show polynomial fit to data.

3.1. Instantaneous and cumulative pressure

The experimental measurements provide pressure estimates that are an average along the line of integration. The FEA models, however, allow us to probe the instantaneous pressure values along a line of integration, and study the phenomenon in greater detail. Simulating the line integration effect for a sound source of known geometry is useful to investigate the practicality of the LDV method for accurate measurements of pressure fields.

First, the instantaneous pressure value along a single scan line (as in Figure 2) was plotted as a function of the distance along that scan line (Figure 4A&C). Then, the cumulative pressure value is calculated along the same scan line (Figure 4B&D). The total cumulative pressure value, defined here as δ, is estimated as $\approx$-750Pa and $\approx$+7,000Pa for point and planar sources, respectively. δ is in effect equivalent to the instantaneous velocity signal from the LDV. We then define the coordinates along the scan line where the cumulative pressure equals 0.95δ (black dots in Figure 4). The 5% is used to reduce the number of data points (black dots), simplifying the analysis without compromising accuracy. Thresholds between 0.9δ and 0.99δ gave similar results, but data were visually less accessible when overlaid onto the pressure distribution.

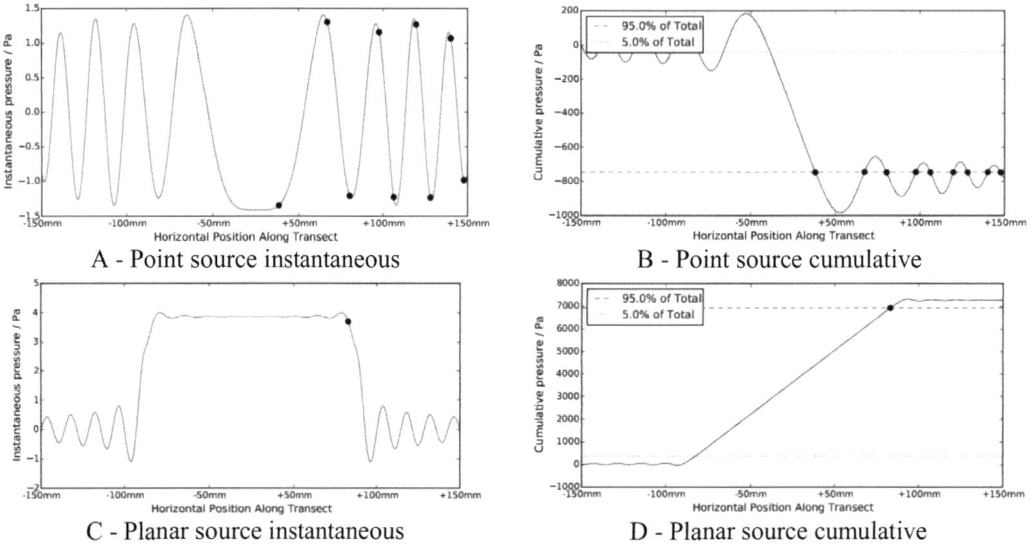

Figure 4. Modelled instantaneous (A-C) and cumulative (B-D) pressure values for a point (A-B) and planar (C-D) radiator, as shown in Figure 2. Black dots indicate locations of equal cumulative pressure, shown only for horizontal position x>0mm.

We then map these coordinates into the instantaneous pressure plots Figure 4A&C. This analytical procedure allows the measurement of the locations along the scan line that contributed to the observed average pressure. There are situations where multiple points reach the average cumulative pressure of 0.95δ. The physical interpretation of this is that the measured average cumulative pressure can be a contribution from several locations along the scan lines, thereby making the effective beam width, a, difficult to establish for a point source. This results in uncertainty about the quantitative interpretation of the LDV signal, and points to the importance of determining the width of the acoustic beam a.

3.2 Acoustic beam width, a

The points in the scanning area that show an instantaneous pressure value equal to 0.95δ were overlaid onto the pressure distribution plot (Figure 5). The sound field was probed using 220 parallel scan lines (corresponding to 15 scan lines per wavelength at 20kHz). This clearly highlights the regions within the scan area where instantaneous pressure contribute to the measured average. The magnitude of the acoustic field can also be computed for different phases; these data are shown in Supplementary Material 1.

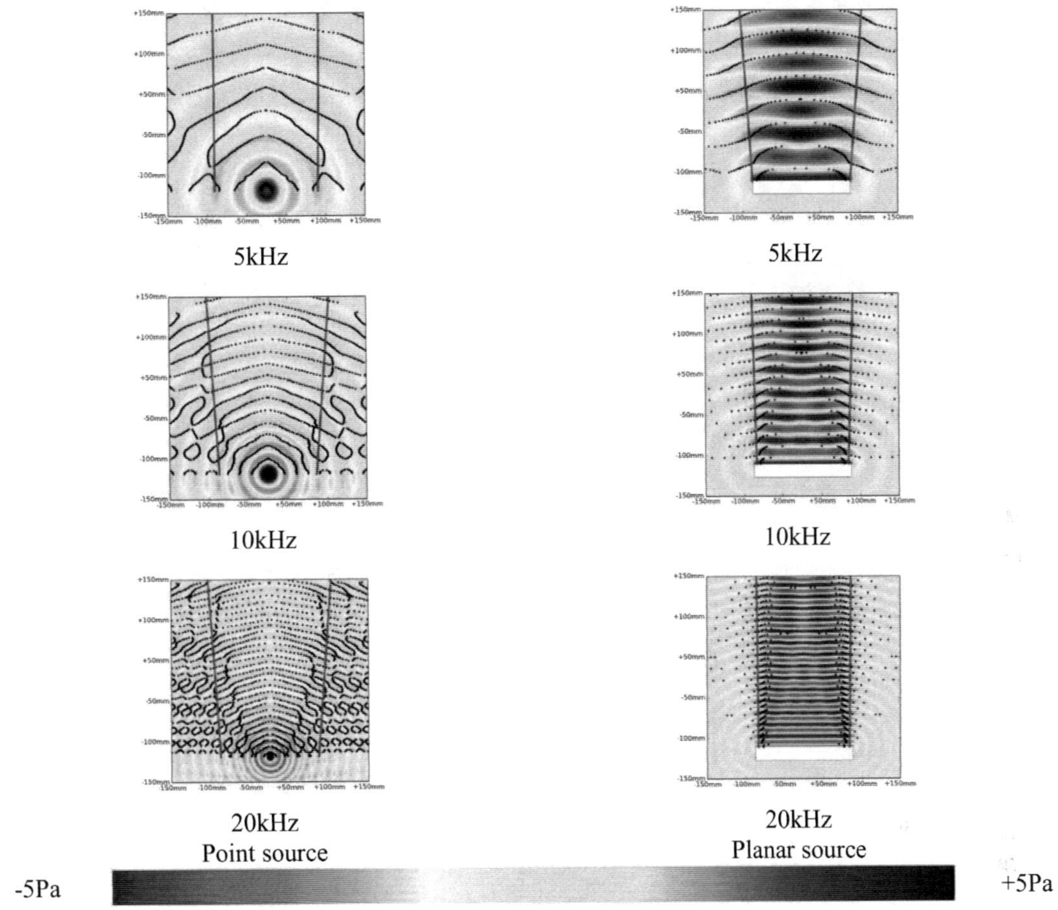

Figure 5. FEA models showing effective beam width for point and planar sources at 5, 10 and 20kHz. Black dots used to estimate beam width as shown in Figure 4. Solid red line shows linear fit to the black points. Red lines are used to calculate the angular divergence (taken as half the angle at the point where both lines meet).

The average width of the acoustic beam, a, was determined by plotting the black dots in a histogram (Figure 6), showing results for 20kHz (histograms for other phases and frequencies are in Supplementary Material 1). The 0.95δ distribution for the planar source yields two clear peaks, revealing the magnitude of beam width a. For the point source producing spherical waves, however, the 0.95δ distribution is also bimodal, yet no clear boundary emerges that permits the unambiguous determination of a.

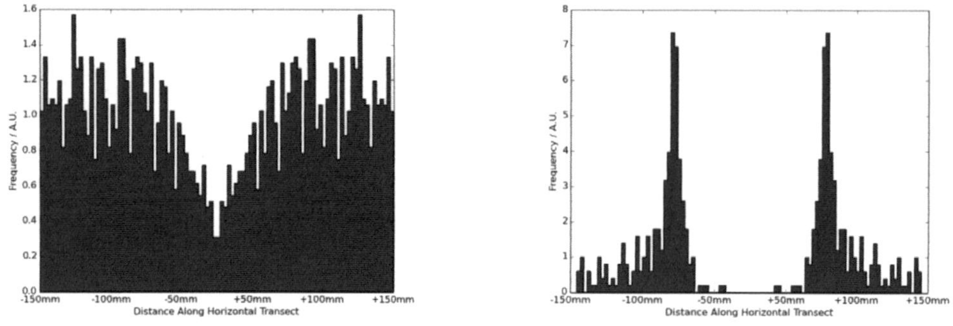

Point source Planar source

Figure 6. Histograms of black dot distributions shown in Figure 5 taken at the same point in time at 20kHz. The point and planar sources clearly differ whereby the latter exhibits two clear peaks with sharp boundaries, defining a.

3.3 Temporal stability and angular divergence of the acoustic beam

The analysis of the sound field at different phase angles allows for the measurement of the angular divergence (defined as half the angle made by the two linear regression lines in Figure 5) of the acoustic beam (Figure 7). The linear regression lines for the point source have very low R^2 values (around 0.002) due to the very large amount of scatter; they are shown here for direct comparison between point and planar sources.

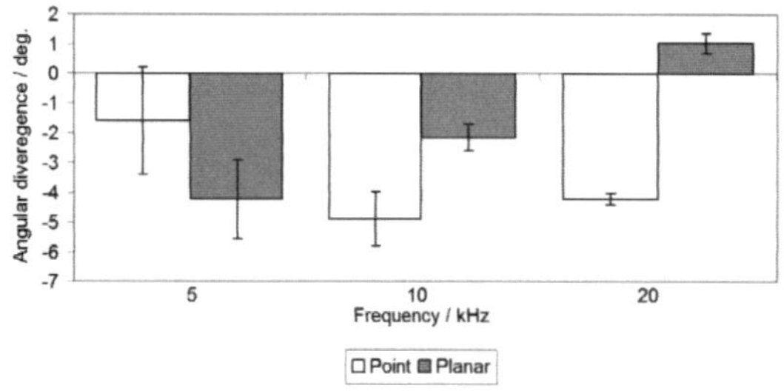

Figure 7 - Calculated angular beam divergence for 5, 10 and 20 kHz. Error bars are ±1 standard deviation, based on divergences measured at various phase angles along the pressure cycle.

The highly scattered black dots for the point source at all tested frequencies (Figure 5 and Figure 6) do not allow for simple estimation of a. For the planar source, the dual peak distribution allows for ready estimation of the beam width. This theoretical analysis provides evidence that a, and its variation calculated at different phase angle can be determined quantitatively using LDV refractometry only for a planar wave front (Figure 8). The analysis also shows the severe limitations encountered for an ideal single point source.

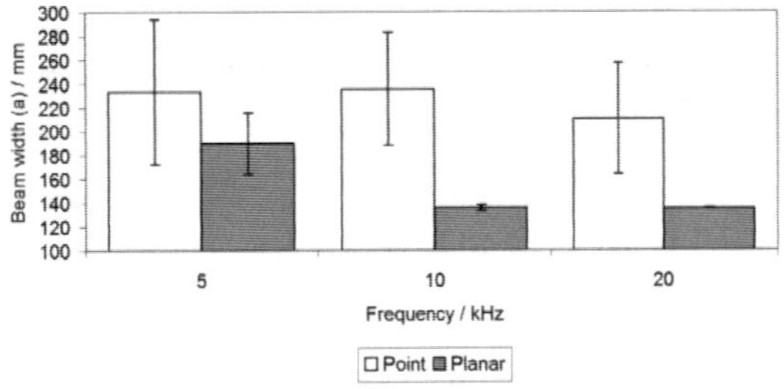

Figure 8. Width of the acoustic beam for different frequencies. Error bars are ± 1 standard deviation, based on measurements at various phases of the cycle.

4. Imaging of the acoustic field

4.1. Experimental procedure

Acoustic reflection and interference were investigated using laser Doppler refractometry, and employing the analytical rationale described above.

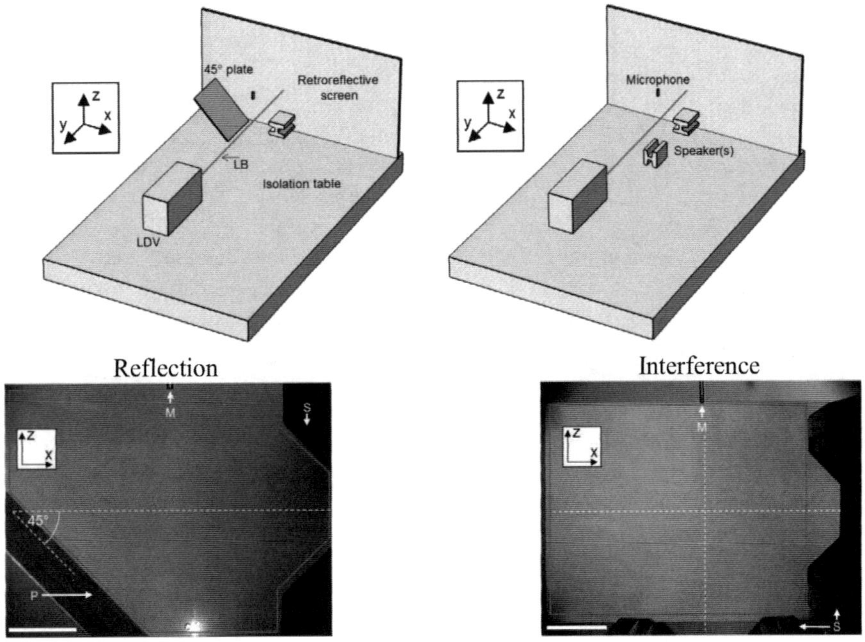

Figure 9. Experimental setup to visualise acoustic reflection and interference. Top panels: Diagrammatic representation of components and their relative positions in experimental setup (not to scale). Red line denotes the path of the laser beam. Bottom panels: Images from LDV showing the scanning area (outlined in red). Scale bars: 50mm. M: microphone, P: plate, S: loudspeaker, LB: laser beam.

The structure and propagation of acoustic waves were visualised for a wave packet of 5 cycles at 20kHz. Other experiments, using frequencies between 5-20kHz and other continuous tones showed similar results. Continuous tones however suffered from noisy signal from the numerous reflections within the acoustic booth used. The 5-cycles wave packet corresponds to 250μs signal duration at 20kHz, a time short enough to allow for repeat measurements (n=50) at each of the scan points, prior to the linear averaging of the velocity signal. The acoustic signal was generated by a waveform generator (Agilent 33120A), and fed to the high frequency planar loudspeakers (AMT-1, ESS laboratories). The sound radiating Mylar membrane of the planar loudspeaker had a rectangular shape of 150mm length and 5mm width. Sampling frequency of the LDV was 512kHz and spatial resolution was set at 15 measuring points per wavelength in air. At 20kHz, this corresponds to ca. 9 points per centimetre. For all measurements, a calibrated precision reference microphone (Brüel & Kjær type 4138) was placed next to the observed pressure field, using a preamplifier (Brüel & Kjær, type 2633), and power supply (Brüel & Kjær, type 5935). The microphone signal is used to provide a calibrated reference in dB sound pressure level (SPL) (re. 20μPa).

One conceivable caveat of the method is that the acoustic energy from the loudspeakers could mechanically actuate the rigid wall used as the reflector. In effect, the laser signal does not distinguish between a fluctuation of the air refractive index along the line of integration and the sound-induced motion of the reflective wall. To investigate whether the recorded signals are affected by such mixing, a loudspeaker was positioned behind the reflective wall and vibrations (which in this case should be purely mechanical) on the reflective wall were measured. At the sound pressures and frequencies used here, the mechanical displacement velocity of the wall did not exceed background noise at all frequencies considered (5-20kHz). This control measurement prompts the conclusion that the reflective wall is indeed rigid during testing and does not contribute any velocity signal to the line of integration.

4.2. Imaging of the acoustic reflection

The reflection experiment reveals the time-resolved evolution of the wave packet as it encounters a reflective aluminium plate of 3mm thickness. Here, a single loudspeaker was positioned such that its acoustic axis was directed towards the aluminium plate at 45° angle of incidence (Figure 9). The temporal evolution of the pressure field was only measured in the space between the loudspeaker and the plate (Figure 10 & Animation 1). Pressure values were calculated from LDV displacement values for each scanning point using equation (5) (with a length of a as determined in 3.3).

For time, $0<t<3.9T$ (T: period at 20kHz), the wave packet is seen to propagate towards the aluminium plate. At $t=3.9T$ the first sign of reflection and interference become apparent. The loudspeaker's inertia results in additional oscillations visible in the wake of the initial wave packet, detectable after $t>1.3$T, and absent in the FE model. At $t = 6.6T$ the wave packet is half reflected from the plate, giving rise to the interference pattern. At $t > 9.2T$ the wave packet has been fully reflected at an angle of 90°.

The experimental data is in good agreement with the finite element simulation, generating near to identical patterns of constructive and destructive interference (Figure 10). LDV and FE simulation pressure field coincide, lending strong support to the claim that the temporally accurate and spatially-resolved visualisation of the acoustic field is possible. There is, however, some disagreement between the measured and predicted pressure values (Figure 10, $t = 0.0T$) very close to the loudspeaker, where the FE model predicts higher pressures than the LDV. This discrepancy is likely due to two distinct effects. First, a small measurement error is introduced from the non-parallel scanning lines. Second, and more importantly, this region of the acoustic field is well within the loudspeaker's near-field, where the structure of sound pressure is notoriously complicated. In fact, little experimental information is available on sound field geometry in the near-field, an outstanding question amenable to the approach presented here.

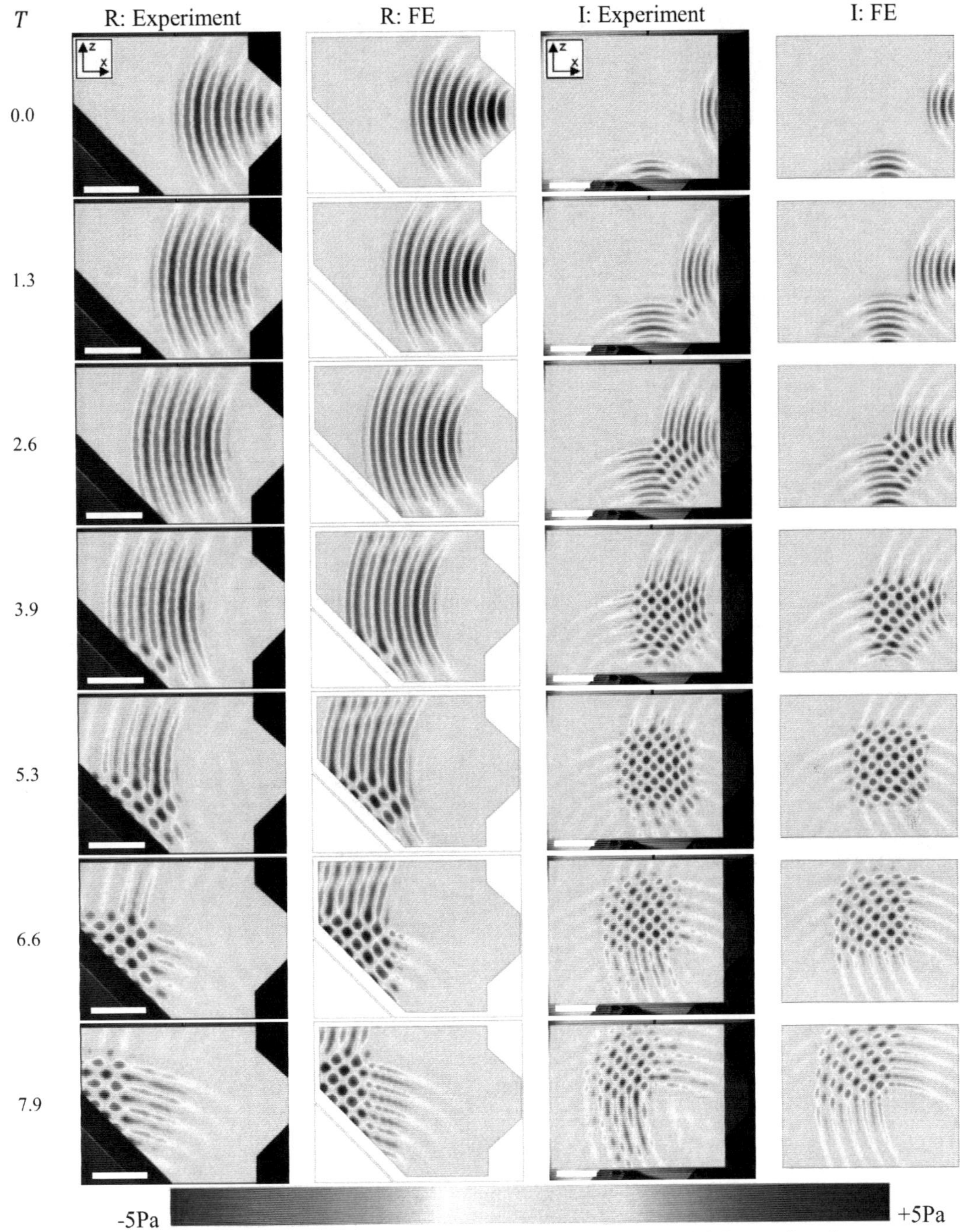

Figure 10a. Experimental and FE representations of pressure field with increasing time steps. Scale bar: 5cm. T: period (T=50μs), R: reflection, I: interference.

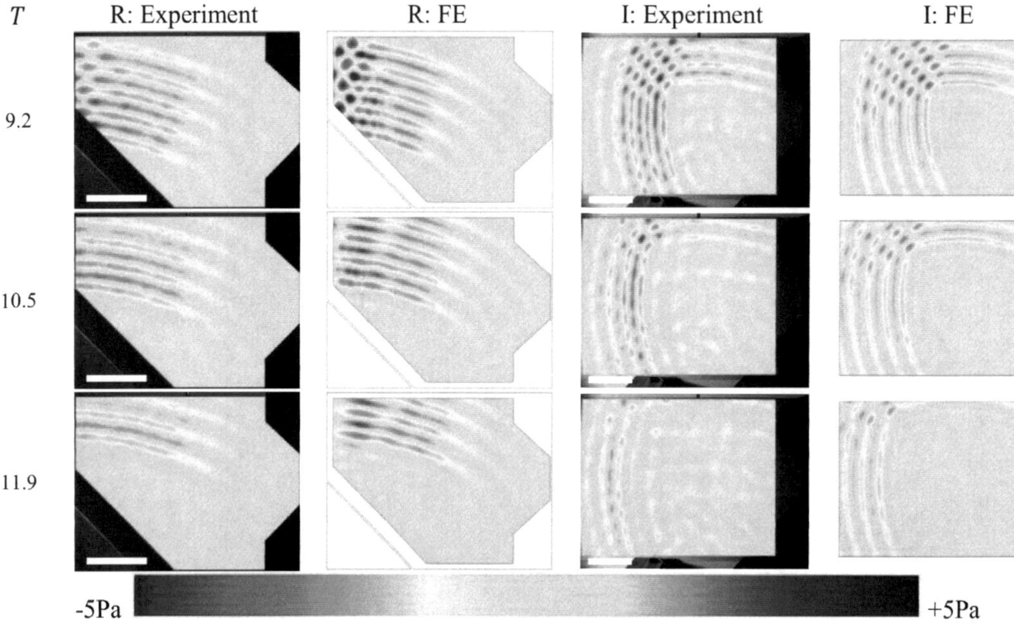

Figure 10b. Experimental and FE representations of pressure field with increasing time steps. Scale bar: 5cm. T: period (T=50μs), R: reflection, I: interference.

4.3. Imaging of the acoustic interference

Acoustic interference was generated using two loudspeakers, oriented at 90° to each. LDV refractometry was then used to image the temporal evolution of the sound field as the two synchronously radiated and identical wave packets interfered (Figure 10 & Animation 2). Interference patterns begin to appear at $t = 1.3T$ and develop fully by $t = 5.3T$. For $t > 5.3T$, the wave packets continue to propagate with only some regions undergoing interference. The structure of the pressure field reveals that some regions are free of interference, where the wave packets propagate independently from each other. Interestingly, the mesh-like region of interference is observed to propagate exactly along a 45° diagonal. For $t > 7.9T$, some afterwaves are observed by the LDV, again resulting from the loudspeakers mechanical inertia. As with the reflection experiment, there is good spatial and temporal agreement between FE simulation and LDV (Figure 10).

4.4. FEA simulation of reflection and interference

Reflection and interference were simulated using a time-dependent acoustic pressure model. A model 2D space of 300mm × 300mm with 15 quadratic triangular elements per wavelength was used. A mesh convergence study was carried out, showing that 6 elements/wavelength would be sufficient to represent the sound field. Yet, at low computational cost, 15 elements were used to match the experimental spatial resolution. The models included an acoustic source boundary to simulate the loudspeaker(s), a sound radiation boundary to simulate the waves propagating out of the model space, and a sound hard boundary to simulate a reflecting surface. The number of elements used was 45k for reflection and 48k for the interference experiments. Results of both simulations are shown in Figure 10. As with most loudspeakers, producing a finite wave packet is difficult to achieve due to the mechanical inertia of the sound radiating elements, therefore some differences in the experimental and FEA results are expected for the trailing edge of the wave packet. As the FEA stimulus is ideal, direct FEA-LDV comparison could be used to investigate the quality of sound radiators in unprecedented temporal and spatial detail.

4.5 Critical evaluation of the method

To verify the quantitative validity of the method, the acoustic pressure measured by LDV was compared to that of the reference microphone. LDV time-resolved signal was taken from a measurement point 1mm away from the microphone (Figure 11). This comparison corroborates previous reports that laser interferometry can be used to infer pressure from refractive index modulation [9] [21]. There is good agreement between microphone and calculated LDV pressure, both in magnitude and phase (Figure 11). Some differences exist between the two signals that are not fully understood. A key difference resides in the fact that the precision microphone detects pressure changes in all three spatial directions, whereas the LDV measurement reports pressure changes along the direction of the laser path, the integration line. In addition, the microphone is detecting sound pressure over a region of air, of unknown volume and geometry. It may therefore be possible that LDV integration done at 1 mm distance does not reflect in detail the pressure integrated by the pressure microphone. To date, information on the actual acoustic sampling of microphones does not provide the spatial details necessary to resolve this issue. The technique presented, applied with an increased scanning resolution, can however be useful to perform such detailed analysis.

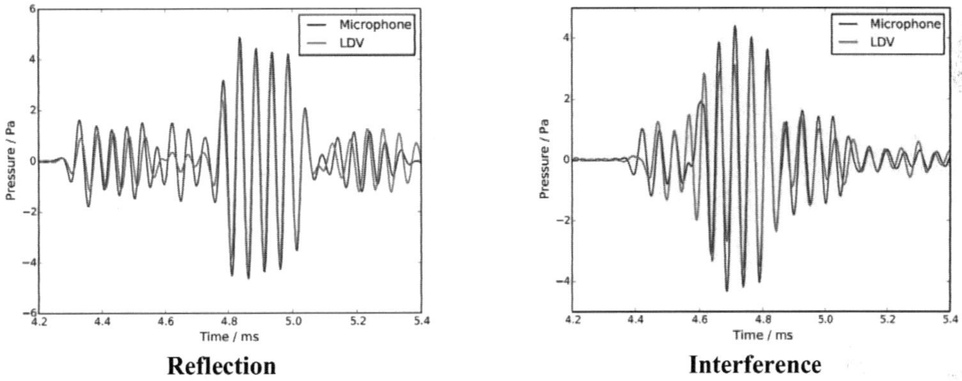

Figure 11 – Comparison of sound pressure for reference microphone and LDV pressure measurements. LDV point taken 1mm from microphone tip.

5. Discussion

We have shown that a quantitative spatial and temporal reconstruction of acoustic waves can be achieved by way of laser Doppler vibrometry and simple transformation of the velocity signal. This study also shows that the geometry of the sound source is an important factor to consider regarding the accuracy of line integration and therefore for the reconstruction of acoustic fields. The measurement error introduced by non-parallel scanning lines can be reduced by a conjunction of three methodological adjustments; the use of a planar loudspeaker with narrow beam width, a sufficiently large laser-reflector distance and suitably large wavelengths. The use of a telecentric lens in the LDV would contribute to the optimisation and flexibility of the technique.

The FEA data (Figure 5) and analysis (Figure 8) together suggest that the effective acoustic beam width of the point source is unstable for the frequencies considered. From Figure 4A it is apparent that multiple points along the integration line contribute to the average signal. The present analysis shows that the quantification of sound pressure radiating from a point source is ambiguous. While the sound field can be visualised, a quantitative approach appears not to be practical at the moment. For the planar source, however, a calculated estimate of the beam width is possible. Whilst the spatial scatter of the beam width, a, for the point source is very large irrespective of frequency (Figure 8), the planar source yields values of a with a clear structure for frequencies >5kHz, revealing a high temporal stability for 20kHz.

Surface reflection and interference hae been accurately reconstructed in terms of spatial organisation, magnitude and phase. FEA corroborates the experimental field visualisation, providing a good match to LDV (Figure 10). The LDV refractometry method also accurately reports on sound pressure level, as confirmed by comparison to a reference precision microphone (Figure 11).

The phenomenological explanation for the line integration phenomenon and the observable pressure field highlights the importance of the geometry and orientation of the integration path. Knowledge of a is required for a correct calculation of the average sound pressure within the observable pressure field, as shown by equation (5). The exact shape of this region is dependent upon the geometry of the acoustic radiator, the distance between the laser beam and the sound radiator, the acoustic interactions with nearby structures, and sound frequency. One consequence is that in sound fields comprising complicated reflective structures, the effective value of a may vary with orientation, setup geometry and sound frequency.

6. Conclusions

The visualisation of sound fields is rarely used as a diagnostic or analytical tool in acoustic research, industrial noise analysis or product development. Refracto-vibrometry is shown here to be a rapid, accurate method that can lend itself to research and industrial applications. The present analysis shows that by carefully considering the geometry of the sound radiator, it is possible to produce accurate sound field imaging. The method presented here does not require extensive calculations and unwarranted assumptions, and therefore could find practical. While some limitations still exist, scanning refracto-vibrometry holds the promise of improving the non-contact, clutter-free quantitative evaluation of acoustic fields.

Acknowledgments

This research is part of a research programme on the mechanics of auditory systems funded by the BBSRC. DR acknowledges support from the Royal Society of London.

References

[1] Chinnery P A 1997 The schlieren image of wo-dimensional ultrasonic fields and cavity resonances *The Journal of the Acoustical Society of America* **101** 250–6

[2] Pitts T A and Greenleaf J F 2000 Three-dimensional optical measurement of instantaneous pressure. *The Journal of the Acoustical Society of America* **108** 2873–83

[3] Nakamura K 2001 Sound field measurement through the acousto-optic effect of air by using laser Doppler velocimeter *4th Pacific Rim Conference on Lasers and Electro-Optics* vol 1 (Chiba-Japan: IEEE) pp 154–5

[4] Zipser L, Franke H, Olsson E, Molin N E and Sjödahl M 2003 Reconstructing two-dimensional acoustic object fields by use of digital phase conjugation of scanning laser

[5] Vuye C, Vanlanduit S and Guillaume P 2009 Accurate estimation of normal incidence absorption coefficients with confidence intervals using a scanning laser Doppler vibrometer *Optics and Lasers in Engineering* **47** 644–50

[6] Jia X, Quentin G and Lassoued M 1993 Optical heterodyne detection of pulsed ultrasonic pressures. *IEEE transactions on ultrasonics, ferroelectrics, and frequency control* **40** 67–9

[7] Olsson E and Tatar K 2006 Sound field determination and projection effects using laser vibrometry *Measurement Science and Technology* **17** 2843–51

[8] Zipser L and Franke H 2007 Refracto-Vibrometry for Visualizing Ultrasound in Gases, Fluids and Condensed Matter *2007 IEEE Ultrasonics Symposium Proceedings* (IEEE) pp 395–8

[9] Torras-Rosell A, Barrera-Figueroa S and Jacobsen F 2012 Sound field reconstruction using acousto-optic tomography. *The Journal of the Acoustical Society of America* **131** 3786–93

[10] Wimberger-Friedl R 1995 The assessment of orientation, stress and density distributions in injection-molded amorphous polymers by optical techniques *Progress in Polymer Science* **20** 369–401

[11] Waxler R M and E W C 1965 Effect of Hydrostatic Pressure on the Refractive Indices of Some Solids *Journal of Research of the National Bureau of Standards -A* **69A** 325–33
[12] Edlén B 1966 The Refractive Index of Air *Metrologia* **2** 71–80
[13] Oikawa Y 2005 Sound Field Measurements Based on Reconstruction from Laser Projections *IEEE International Conference on Acoustics, Speech, and Signal Processing* vol 4 (Philadelphia-USA: IEEE) pp 661–4
[14] Zipser L, Seelig H-D and Franke H 2009 Refracto-vibrometry for visualizing ultrasound in small-sized channels, cavities and objects *International Ultrasonics Symposium* (Rome-Italy) pp 2588–91
[15] Olsson E 2009 Three-dimensional selective imaging of sound sources *Optical Engineering* **48** 035801
[16] Wang Y, Theobald P, Tyrer J and Lepper P 2004 The application of scanning vibrometer in mapping ultrasound fields *Journal of Physics: Conference Series* **1** 167–73
[17] Nakamura K, Hirayama M and Ueha S 2002 Measurements of air-borne ultrasound by detecting the modulation in optical refractive index of air *IEEE Ultrasonics Symposium* (Munich-Germany) pp 609–12
[18] Dittmar A and Behrendt R 2008 Measuring the 3D propagation of sound waves using scanning laser vibrometry *Berlin Beamforming Conference* (Berlin)
[19] Olsson E 2007 Selective imaging of sound sources in air using phase-calibrated multiwavelength digital holographic reconstructions *Optical Engineering* **46** 075801
[20] Solodov I, Döring D and Busse G 2008 Air-coupled laser vibrometry: analysis and applications *Applied Optics* **48** 33–7
[21] Balek R and Slegrova Z 2002 A Comparison of Acoustic Field Measurement by a Microphone and by an Optical Interferometric Probe *Acta Polytechnica* **42** 13–7

Supplementary Material 1

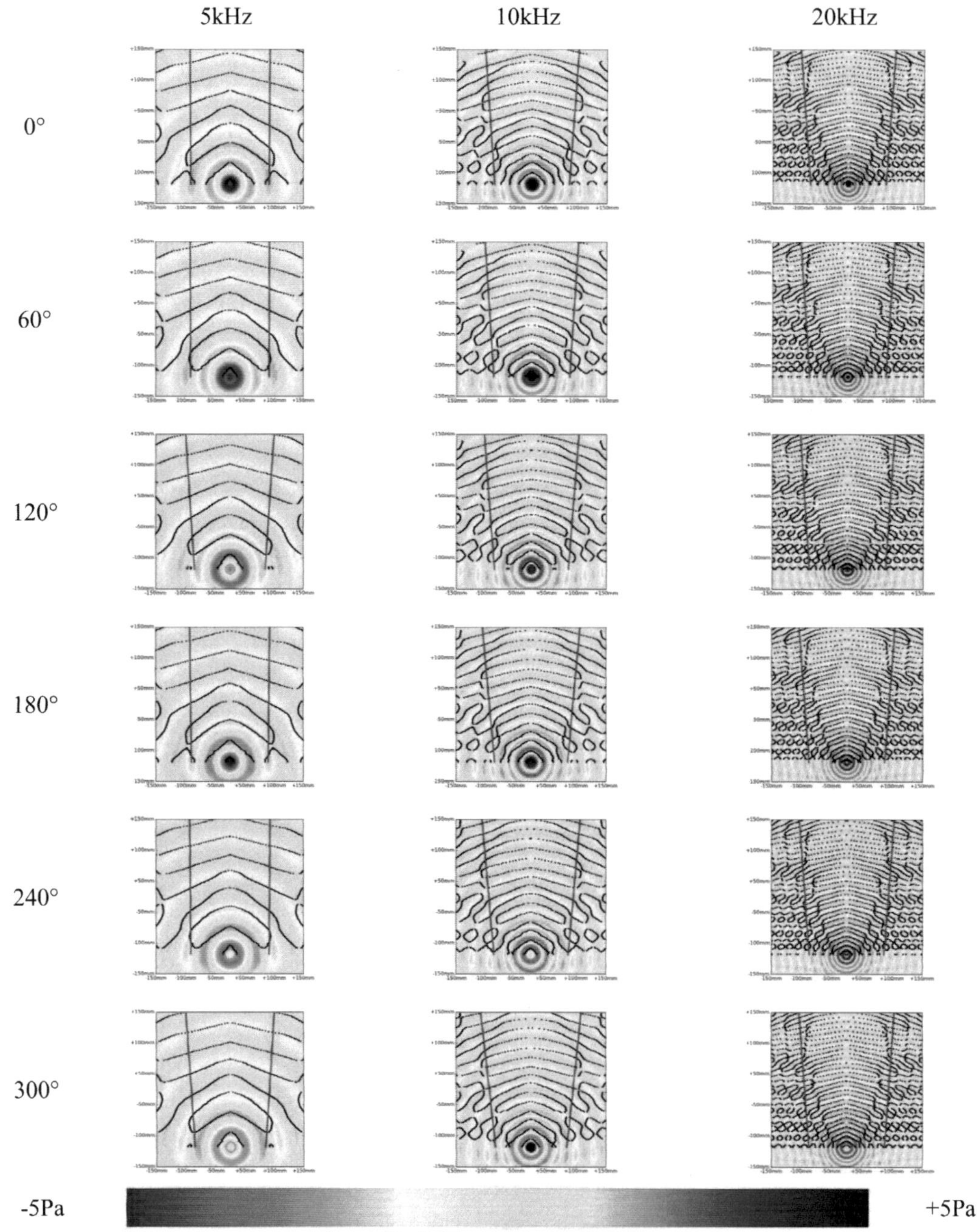

Figure S1 - Point source temporal stability.

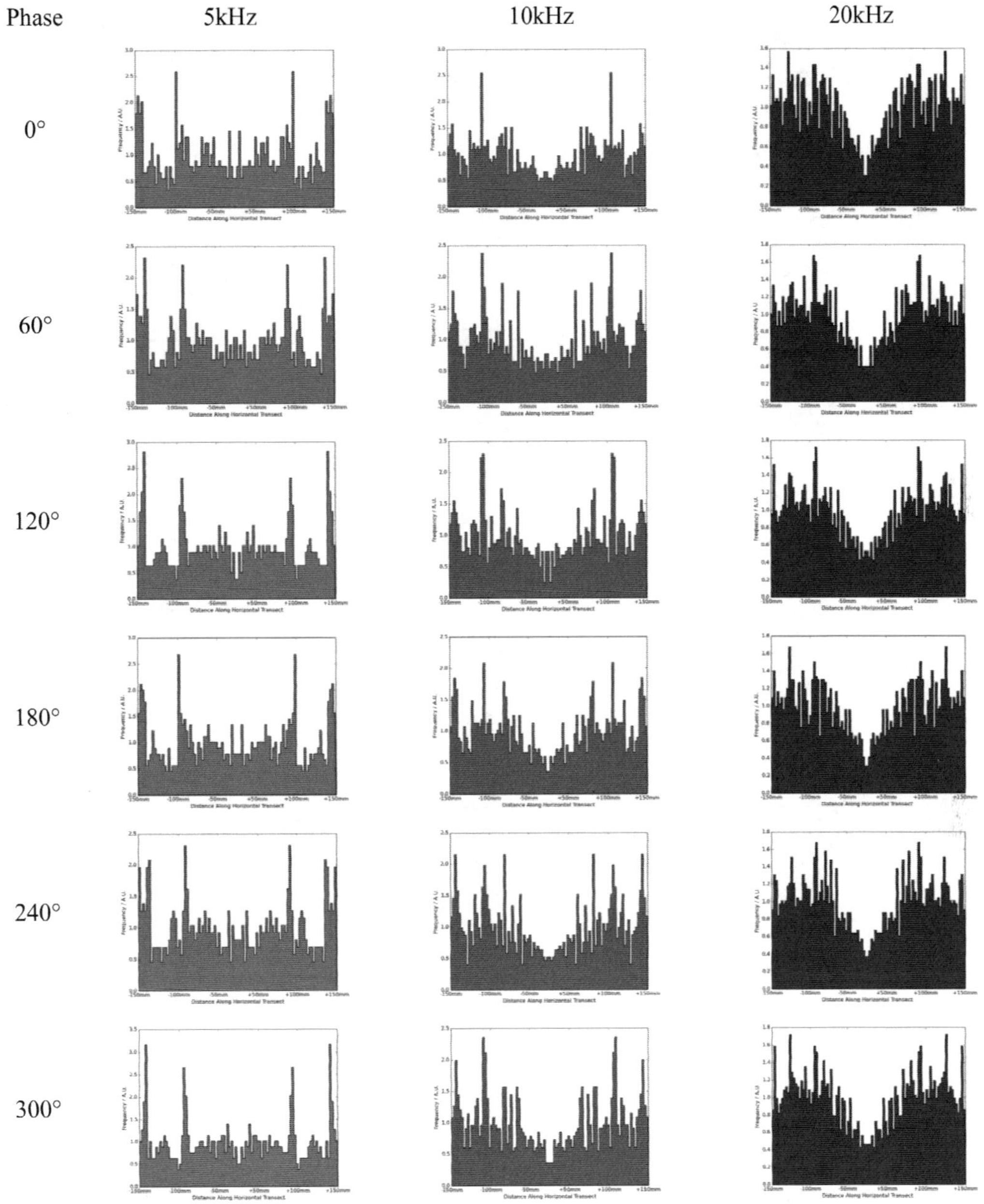

Figure S2 - Point source histograms.

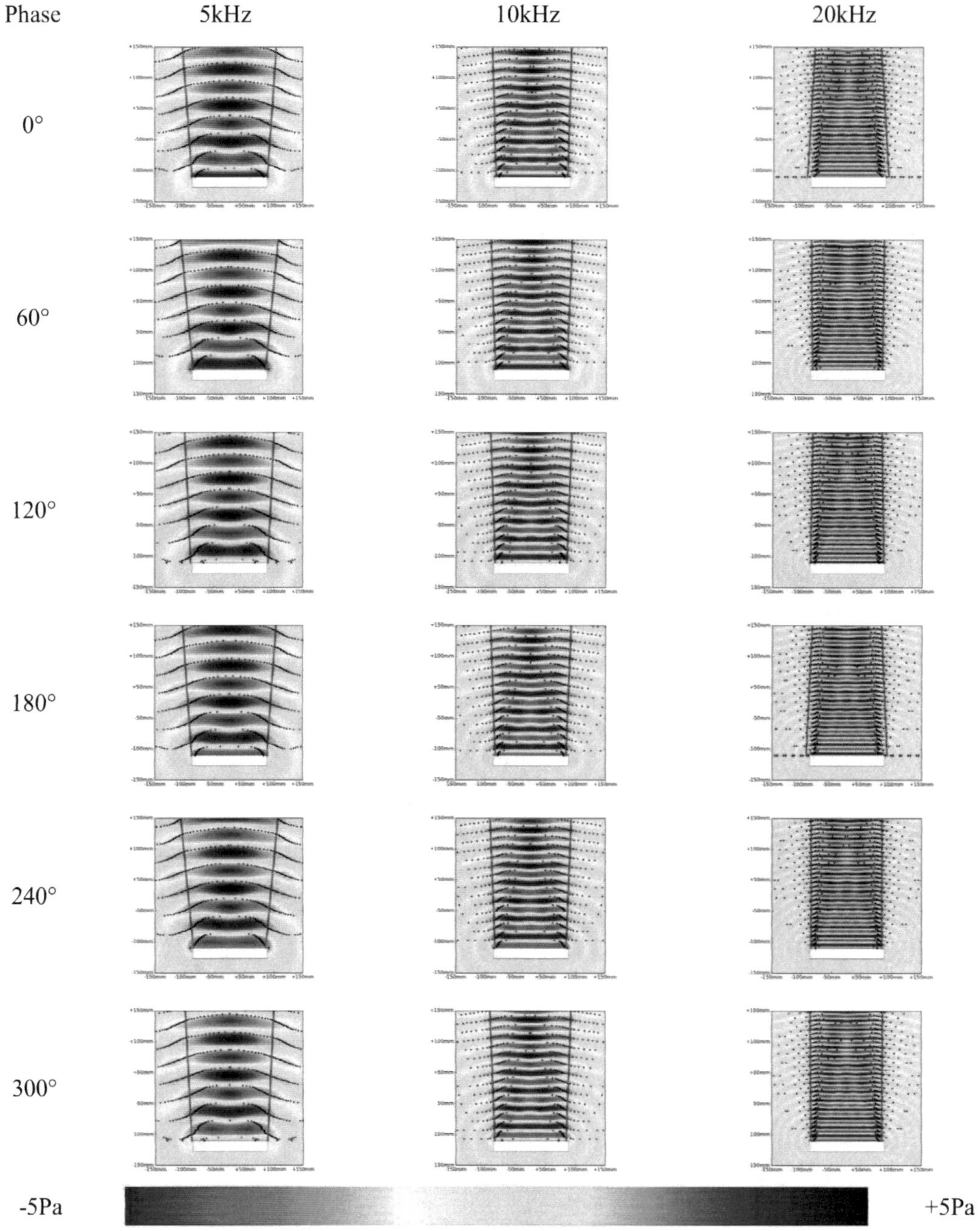

Figure S3- Planar source temporal stability.

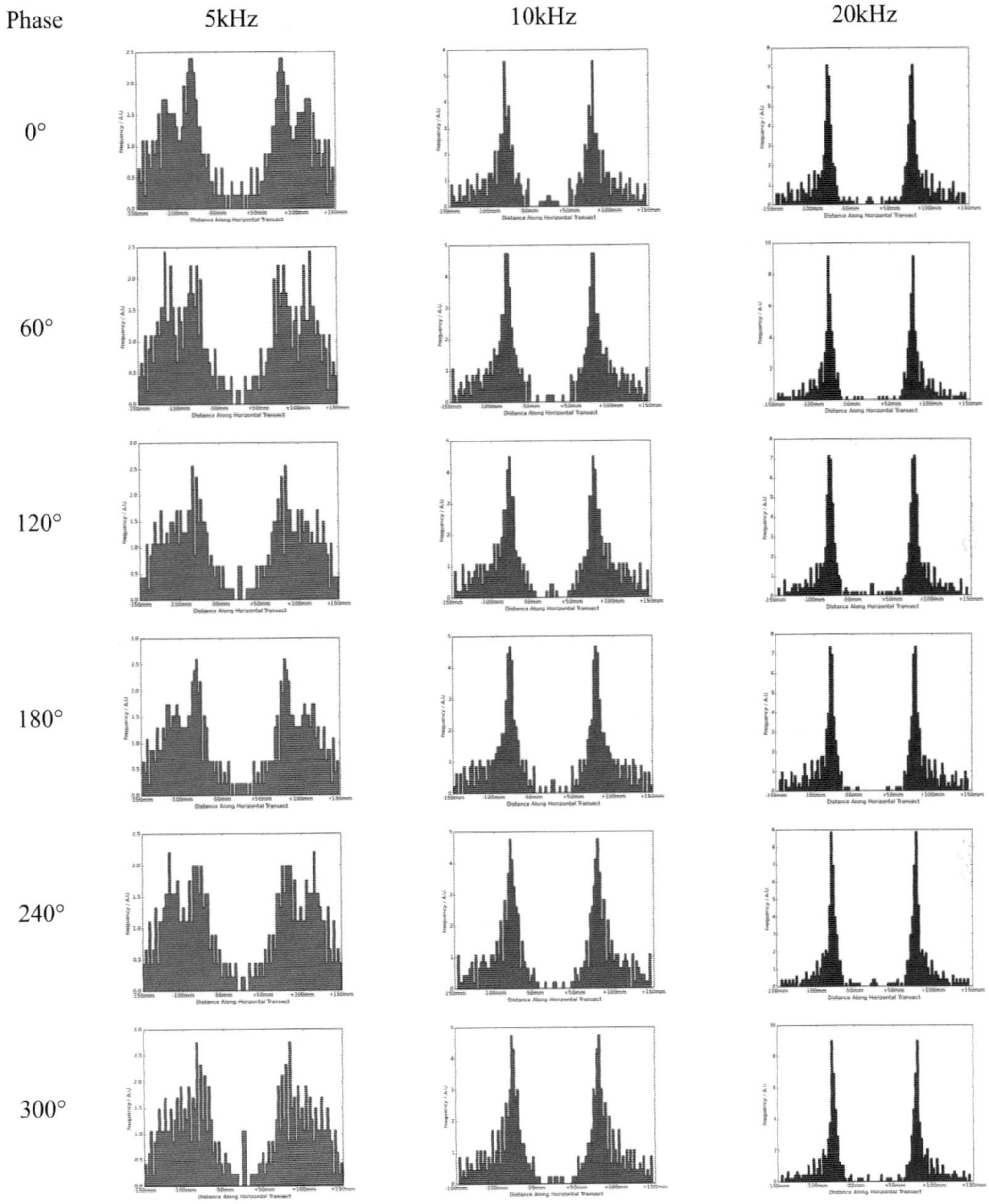

Figure S4 - Planar source histograms.

COMSOL modelling of the acoustoelastic effect

N J Watson[1], T Hazlehurst[1], M J W Povey[1], A Drennan[2], P Seaman[2]

[1]School of Food Science and Nutrition, University of Leeds, UK
[2]Strainsonics Ltd, Sheffield, UK

E-mail: N.J.Watson@leeds.ac.uk

Abstract. Many structural components are subjected to either constant or temporal mechanical loads, such as a suspension bridge bolts and rail tracks. Methods are required to accurately and efficiently measure the stresses experienced by these components to ensure they can continue to operate in an effective and safe manner. Acoustic techniques can be used to monitor the stress in a solid material via the acoustoelastic effect. This is the stress dependence of the acoustic velocity in an elastic media. This work develops a multiphysics computational model to study the acoustoelastic effect in a three point bending system. A simple linear relationship was utilised to represent the stress effect on the acoustic velocity. The simulation results were compared with experimental results and the same general trend was observed. An increase in applied load resulted in a greater difference between the time of flight of two transducers at the top and bottom of a component and perpendicular to the applied load. However, there were quantitative differences between the model and the experiment. The model was used to investigate different ultrasound transducer location and operating frequency, highlighting the benefit of modelling tools for the design of acoustic equipment.

1. Introduction

The acoustoelastic effect is the stress dependence of an acoustic wave's velocity propagating in an elastic media. By measuring the speed of sound in solid components it is therefore possible to calculate the stresses they are experiencing. This can be used for nondestructive testing on many components such as load bearing bolts on suspension bridges [1], concrete materials [2] steel pipes [3] and rail track [4].

The speed at which a wave propagates within a medium is a function of the density and elasticity of a medium. As stress results in changes in these properties it will have an effect on the propagating wave's speed. The speed of sound in a material can be calculated based on the second and third order elastic constants. When a longitudinal acoustic wave is travelling parallel to the stress in a material:

$$\rho_0 c_{Lx}^2 = \lambda + 2\mu - \frac{\sigma}{3K_0}\left[2l + \lambda + \frac{\lambda+\mu}{\mu}(4m + 4\lambda + 10\mu)\right] \qquad (1)$$

Where ρ_0 is the initial density, K_0 is the initial bulk modulus, c_{Lx} is the longitudinal speed of sound parallel to the applied stress δ. λ and μ are the Lamé or second order elastic constants and l and m are Murnaghan's or third order elastic constants. There is an additional third order elastic constant n which is required for propagating shear waves.

When the longitudinal wave is travelling perpendicular to the stress this equation becomes:

$$\rho_0 c_{Ly}^2 = \lambda + 2\mu - \frac{\sigma}{3K_0}\left[2l + \lambda + \frac{2\lambda}{\mu}(m + \lambda + 2\mu)\right] \quad (2)$$

Where c_{Ly} is the longitudinal speed of sound perpendicular to the stress. Often for simplicity a single acoustoelastic constant K (which is a function of the third order elastic constants) is defined. A different K must be given for each type of wave and direction of propagation relative to the stress. The resulting modulation on the speed of sound in the material can then be found:

$$K\sigma = \frac{\Delta C}{C} \quad (3)$$

Where σ is the stress C is the speed of sound in the unstressed material and ΔC is the change in speed of sound. Acoustoelastic constants can be calculated by subjecting materials to different stresses and measuring the speed of sound in those materials. Many components under stress in the real world do not experience either tension or compression in a single direction; therefore methods are required to study them. For example many components experience a three point bending load where there are different regions under either tension or compression.

Computational modelling is a powerful tool when developing ultrasound instruments. They can aid in the design of the system by studying the effect of transducer parameters such as frequency and dimensions and aid with identifying the most suitable location for the transducers. Computational models have been used to study the acoustoelastic effect such as Ferrari who developed a model for wave propagation in metal sheets using EMATs [5] and Elyea et al who studied the effect of bridge bolt dimensions on wave propagation in the presence of tensile stress [6].

In this work we use COMSOL multiphysics to combine structural deformation with acoustic wave propagation to create a 2D model of a component experiencing a three point bending stress. This is representative of a load bearing bolt used in suspensions bridges. Experimental acoustic measurements were made on a bolt and compared to the simulation results. The model was then used to investigate transducer frequency and position as well as the effect that the applied load has on the wave velocity and attenuation.

2. Methodology

2.1. Model development

The computational model was developed using COMSOL multiphysics 4.3. The model combined the structural mechanics and acoustics modules. A 2D model was used for this study as the large number of mesh elements required would result in an unacceptable solution time if a 3D model was used.

2.1.1. Geometry. The geometry of the model can be seen in Figure 1. The rectangle has the same length as the bolt used for the experiments and the height is the same as the diameter of the bolt. The two points in the dashed ellipses represent either end of a line where an oscillating pressure is applied. These represent the ultrasound transducers and have a length of 4 mm. This represents the diameter of 4 mm, which was the same dimension as the transducers used for the experiment.

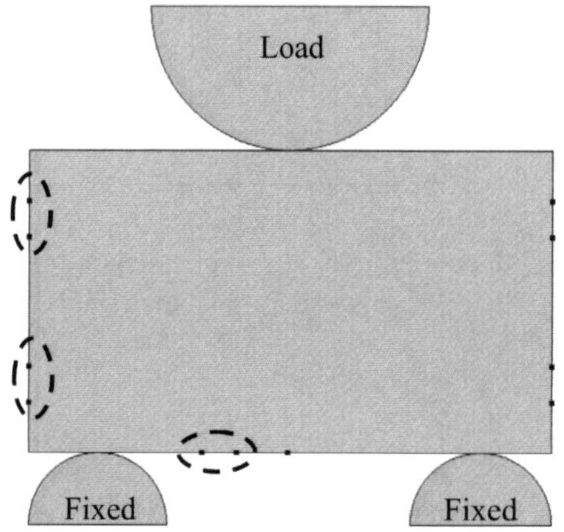

Figure 1. Model geometry with load constraints and transducer points.

2.1.2. Structural mechanics. The rectangular representation of the bolt was placed on two hemispheres which were fixed in both directions. The bottom surface of the rectangle was allowed to slide along these surfaces as deformation occurred. The load was applied to the top surface of another hemisphere which was in contact with the top surface of the rectangle. The material for all of the components in the model was set as structural steel. For the simulations different forces were applied, ranging between 50 and 300 kN.

2.1.3. Acoustic wave propagation. The acoustic wave was added to the system by applying a pressure wave to the lines which represent a transducer. An analytical function was defined which represented a shaped pulse with known frequency and duration. The amplitude was always set to 1 in the positive direction. For the majority of the simulations a frequency of 1 MHz was used. This is much lower than the 10 MHz transducers used for the experiments. This reduction was required as the frequency controls the element mesh size and simulation time step. A frequency of 10 MHz would have resulted in an unacceptably long time to run the simulations. The acoustoelastic effect was modelled using equation 3 above.

For all simulations the value for the acoustoelastic constant K was selected as 5e-7. Methods do exist for measuring the acoustoelastic constant for specific materials, load and propagation directions [7]. Unfortunately the scope of this work did not allow for the exact measurement to be made at his time. The system was operated in a pulse echo mode so the transmitting transducer was also the receiver. This resulted in an acoustic path length which is twice the length of the bolt.

2.1.4. Finite element mesh. The geometry of the system was discretised using a finite element mesh. A free triangular mesh was used for all models. An important aspect is the size and number of mesh elements utilised as this governs the accuracy, precision and time duration of the simulation. A trade off always exists to this extent. Adopting the method of Mylavarapu and Boddapati [8] the maximum element size used was the ultrasound pulse wavelength divided by ten. This results in ten elements for each cycle of the waveform. The wavelength was calculated from the speed of sound and the transmitted frequency used.

2.1.5. Solver configurations. All simulations had two steps for the solver. The first step was a stationary solver and the second step was a time dependent solver. The stationary step was used to solve the solid mechanics and the results from this (stress) were then used during the second step, which solved the acoustic propagation. The applied load would result in a deformation to the bolt so the resulting deformed mesh was utilised during the second step to account for the changing path lengths the acoustic waves would now experience. The selection of the simulation time step is important for time dependent studies as it controls the accuracy resolution and simulation duration. Mylavarapu and Boddapati [8] stated that the simulation time step should be less than ((wavelength/10)/speed of sound) to ensure mathematical stability. This approach was used in this work.

2.1.6. Data analysis. The results taken from the simulations were the time of flight and the amplitude of the received acoustic wave. The pressure received at the line representing the transducer was recorded at each time step and exported to MATLAB for post processing. The time of flight was calculated by detecting the time at which the first part of the received pulse over an amplitude value of 0.1 was detected. The amplitude recorded was the maximum amplitude of the received acoustic wave.

2.2. Experimental measurements
The experimental trials were all performed at Strainsonics Ltd, Sheffield, UK. The bolt was placed in a fork and pin connector (Figure 2). This connector was then placed in a load frame and the required load was applied. The bolt has two 10 MHz ultrasound transducers, each with a diameter of 4 mm built into one of the bolt ends. The bolt was orientated so that the transducers were placed at the top and bottom of the bolt both parallel to the applied load. The transducers were attached to a T200 ultrasound pulser receiver device (Tribonsonics, Sheffield, UK). The time of flight was recorded for the top and bottom transducers, for applied loads between 50 and 300 kN.

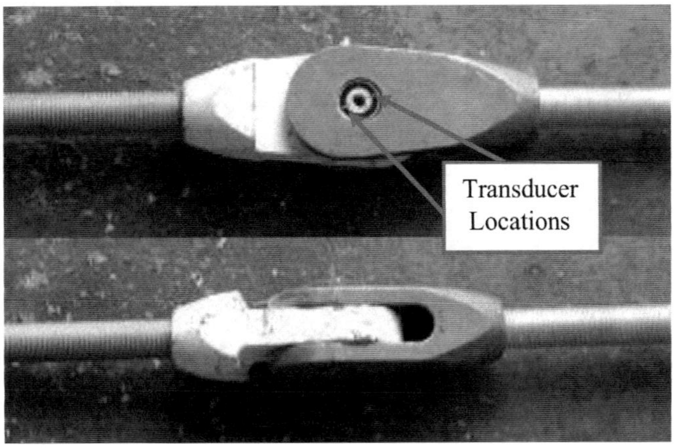

Figure 2. Fork and Spade connection for experiments (Courtesy of Strainsonics Ltd). The bolt with the built in ultrasound transducers is located in the center of the connection.

3. Results and discussion

The simulation results show that the time of flight is lower for the top transducer indicating a higher speed of sound in this region (Figure 3). This was expected as the top zone was closer to the applied load and a larger stress would be experienced here. This region is experiencing a compressive stress and ultrasound velocity is known to increase with compressive stress [1]. The top section of the bolt is also under compression so the acoustic path length is shortened here, again reducing the time of flight of the ultrasound pulse. The results indicate that an increase in load results in a shorter time of flight for the top transducer and a longer time of flight for the bottom transducer. Again this is an expected result as the increased load will result in an increase in the stress which is modulating the speed of sound in the material. The results for the loads of 100 and 150 kN do not seem to fit with the rest of the results as the time of flight for both transducers are much higher. It is believed that this is a result of how the time of flight is measured. The time of flight is taken as the point at which the received ultrasound pulse crosses a threshold value of 0.1. In some cases it was noticed that the first cycle of the received pulse was too low to break this threshold so it would be the second cycle from which the time of flight was calculated. This could have been the case for the results for 100 and 150 kN, resulting in this abnormality. The sudden jump in time of flight is of the order of the time period of the wave again supporting this explanation. These errors could be removed by altering the threshold level during the post processing of the results.

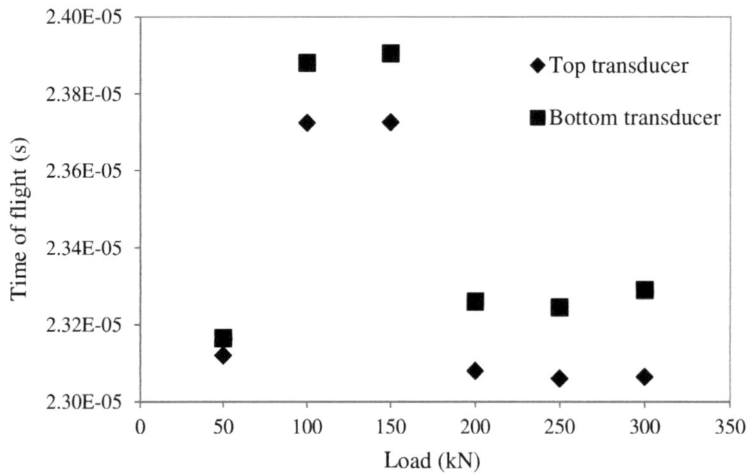

Figure 3. Simulated ultrasound time of flight for top and bottom horizontal transducers, as function of load.

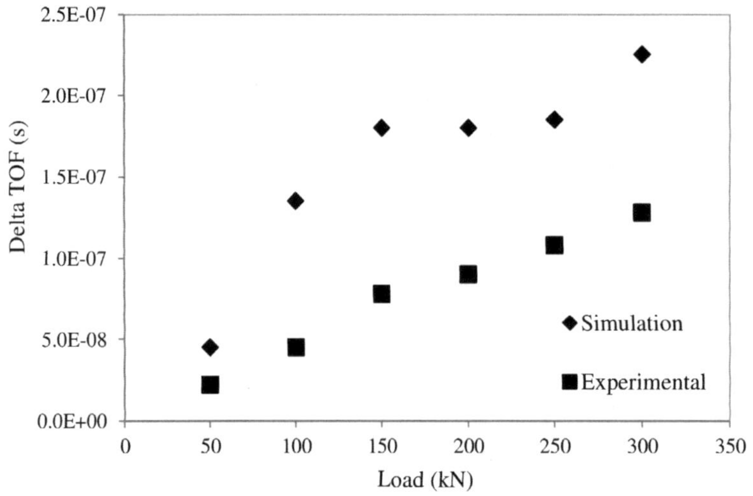

Figure 4. Difference in time of flight for the top and bottom horizontal transducers. Results are presented from the simulation and experiment.

Figure 4 display the difference in acoustic time of flight between the top and bottom transducers in the bolt from the simulation and experiment. Both sets of results show that an increase in load results in an increase in the difference in time of flight. This result indicates that a simple linear modeling technique is capable of recreating some features of the acoustoelastic effect. The simulation results are higher for all load values than the experimental results. The experimental results display an almost linear relationship between load and delta time of flight. For the simulation results this is almost the case except for the values of 150 and 200 kN. For these values there seem to be no change in delta time of flight. This is most likely a result of how the time of flight is calculated as discussed in the previous paragraph. It was also noticed during the simulations that the propagating waves from the top and bottom transducers would overlap each other and interfere (Figure 5). This would introduce some error into the results. This would not happen for the experiments as the transducer frequency was much higher resulting in much less beam spread. When simulations were performed with higher frequencies this interference was reduced.

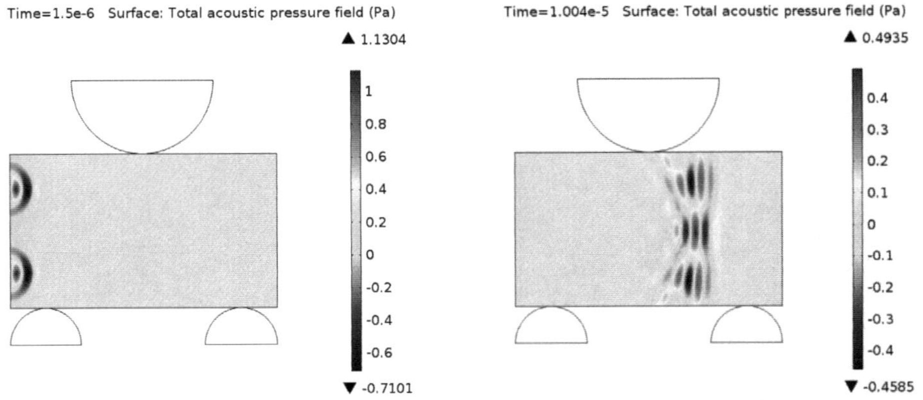

Figure 5. Acoustic pressure field in the system: Left, time = 1.5e-6 (s), right, time = 1.004e-5 s.

The results indicate quantitative differences between the models and the experimental results. It should be noted that the model used to calculate the speed of sound as a function of stress is a simple linear equation. These equations use a single value for the acoustoelastic constant *K*. A simplification to a single value is only valid when the stress is acting in a single direction. For the model we created a three point bending system. This type of structural deformation has regions of tension and compression so a single value for the acoustoelastic constant is not valid and the equation should be further developed to include all three of the third order elastic constants. It should also be remembered that the simulation was a 2D model whereas the experiment was a full 3D system; again this would result in differences between the simulations and experiments.

The amplitude ratio was calculated by the following equation:

$$AR = \frac{A_{top} - A_{bottom}}{A_{top}}$$

Where A_{top} is the maximum amplitude of the received pulse from the top transducer. A_{bottom} is the maximum amplitude of the received pulse from the bottom transducer. This equation results in a ratio of the received amplitudes relative to the top transducers amplitude. It was decided to make the ratio relative to the top transducers received amplitude as this was always found to be higher in the simulation results. The results in Figure 6 indicate that an increase in applied load results in an increase in the difference in the received pulse amplitude. This relationship appears to be almost linear with the only outlying result the one for a load of 100 kN. This could be due to the interfering waves. The amplitude of the received pulses was not recorded from the experiments so a direct comparison with them cannot be made. These simulation results indicate that in addition to measuring the time of flight and calculating the speed of sound in a material experiencing a stress, amplitude measurements may also be a viable method for measuring stress provided that appropriate models are developed. Using different measurements would increase the versatility of an ultrasound inspection system.

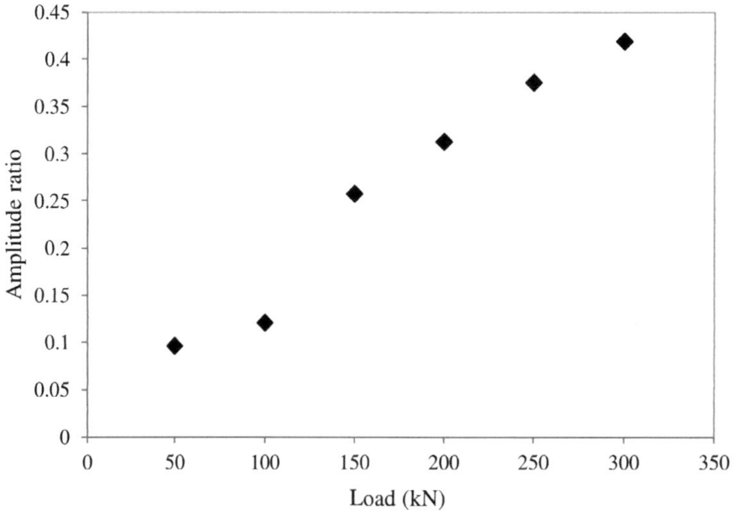

Figure 6. Simulated amplitude ratio of received acoustic waves from the top and bottom horizontal transducers, as a function of load. The acoustic waves had a frequency of 1 MHZ.

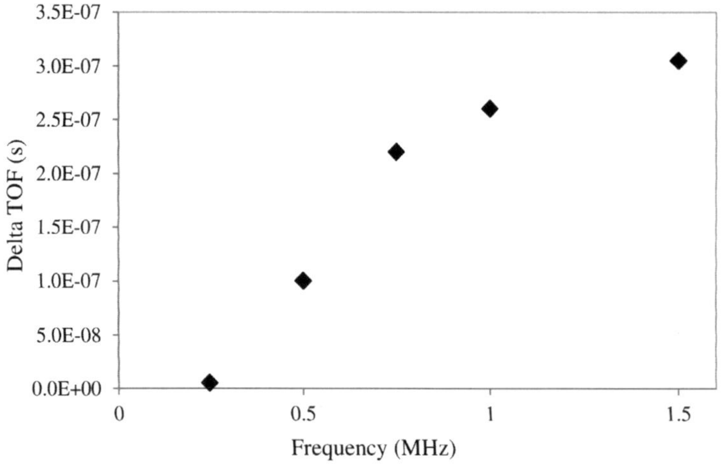

Figure 7. The effect of acoustic wave frequency on the change in time of flight for the top and bottom horizontal transducer. Simulation results at a load of 150 kN.

The frequency of the transmitted acoustic wave was varied for some simulations (Figure 7). The frequency was varied from 0.25 to 1.5 MHz. No frequency was simulated above this value due to the required increase in number of mesh elements and the reduction in simulation time step resulting in too long a simulation duration with the current computational capabilities. The Simulation results show that the higher the acoustic wave frequency the greater the difference in time of flight. This indicates that higher frequencies are more suitable for detecting stress and for identifying smaller changes in stress. High frequency waves will experience much less beam spread reducing the chance of interference and sampling a smaller area of stress. It should always be remembered that attenuation does increase with frequency and a trade-off always exists with this respect. The result for the effect of frequency on the amplitude ratio again showed that higher frequencies are more preferable as the different was greater for the higher frequencies (Figure 8). This relationship almost appears to be linear.

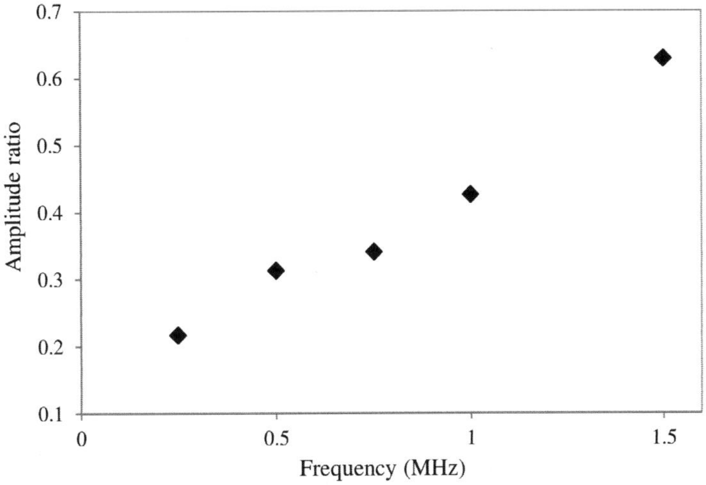

Figure 8. Simulated amplitude ratio of received acoustic waves from the top and bottom horizontal transducers, for different acoustic wave frequencies. The applied load was 150 kN.

One benefit of using computational modelling when developing ultrasound instruments is to investigate the location of the transducers. This is done easily in a model where it only requires the location of the oscillating boundary to be moved. Transducer locations can be tested automatically with parameterised solvers to find the precise optimum location. In addition to the two horizontal transducers at the top and bottom of the bolt another transducer location was defined along the bottom edge with a wave travelling parallel to the applied load (Figure 1). It should be noted that for the model where the wave is propagating in the same direction as the applied load for convenience the same value for the acoustoelastic constant was used. This would not be the case in real life.

The results for the time of flight indicate that a larger applied load result in a longer time of flight (Figure 9). This result is a good linear fit. This is unlike the previous results where measurements were made based on two propagating wave. This potentially indicates the observed abnormalities were a result of the two waves interfering. The wave travelling in the vertical direction experiences regions of both compression and tension so it is interesting to observe the overall effect on the time of flight.

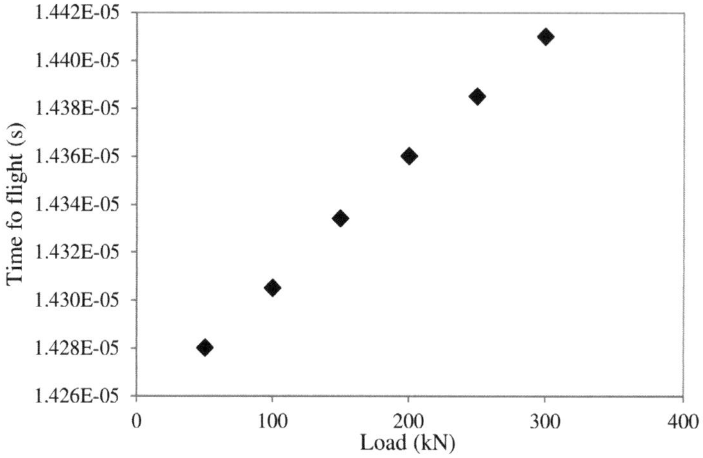

Figure 9. Simulated time of flight for the vertical transducer, as a function of load.

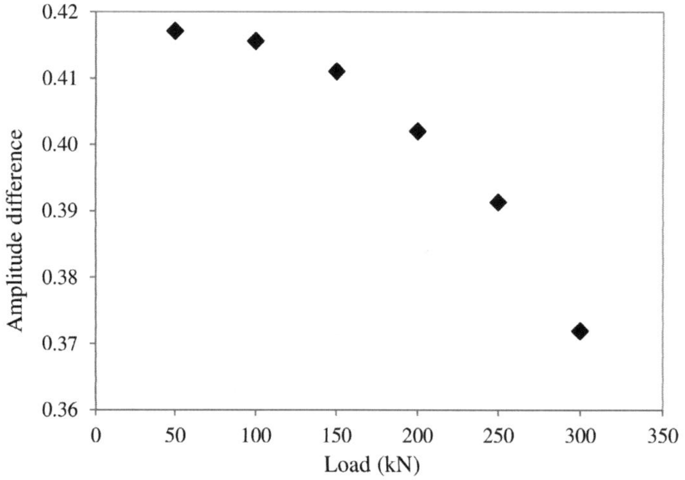

Figure 10. Simulated amplitude difference for the vertical transducer, as a function of load.

The amplitude difference for the single vertical transducer was calculated by deducting the maximum positive amplitude of the received pulse from 1, which was the transmitted pulse positive amplitude. The results for the amplitude difference as a function of load indicates that with an increase in load the difference in amplitude between the transmitted and received pulse reduces (Figure 10). This relationship is not linear and shows very little change between 50 and 150 kN before it reduces with increased load. Throughout the range of loads applied the change is only small (0.05) suggesting that when the ultrasound propagation in parallel to the applied load, the change is not great.

4. Conclusion

This work has developed a multiphysics model combining structural analysis with acoustic wave propagation to study the acoustoelastic effect in a three point bending system. Although a greatly simplified model was used with a linear relationship between applied stresses and speed of sound the key features of the acoustoelastic effect were simulated and qualitatively the simulation results shared the key features of experiment. This model showed that increasing the load applied to the system would result in a greater difference in time of flight and received acoustic wave amplitude for two transducers in different location but both perpendicular to the applied load. The effect of transmitted wave frequency and transducer orientation was also studied. The results indicate modeling techniques such as this can be powerful design tools. Further work would require more detailed modelling of the acoustoelastic effect at higher frequencies and in three dimensions.

5. Acknowledgements

The authors would like to acknowledge Nanofactory and the European Regional Development Fund for their financial support of this research.

References

[1] Si-Chaib M O, Menad S, Djelouah H, Bocquet M 2001 *NDT&E International* **34** 521-529
[2] Lillamand I, Chaix J F, Ploix M A, Garnier V 2010 *NDT&E International* **43** 655-660
[3] Javadi Y, Pirzaman H S, Raeis M H, Najafabadi M A 2013 *Materials and Design* **49** 591-601
[4] Chen F, Wilcox P D 2007 *Ultrasonics* **47** 111-122
[5] Ferrari M G 2009 Proceedings of the 2009 COMSOL Conference in Milan
[6] Elyea A, Doubek B, Hubbard G, Ozevin, D 2011 Proceedings of the 2011 COMSOL Conference in Boston
[7] Egle D M, Bray D E 1976 *The Journal of the Acoustical Society of America* **60** 741-744
[8] Mylavarapu P S K and Boddapati S 2011 Proceedings of the 2011 COMSOL Conference in Bangalore

Optical detection of ultrasound from optically rough surfaces using a custom CMOS sensor

S O Achamfuo-Yeboah, R A Light and S D Sharples

Applied Optics Group, University of Nottingham, Nottingham, United Kingdom, NG7 2RD

E-mail: `samuel.achamfuo-yeboah@nottingham.ac.uk`

Abstract. The optical detection of ultrasound from optically rough surfaces is severely limited when using a conventional interferometric or optical beam deflection (OBD) setup because the detected light is speckled. This means that complicated and expensive setups are required to detect ultrasound optically on rough surfaces. We present a CMOS integrated circuit that can detect laser ultrasound in the presence of speckle. The detector circuit is based on the simple knife edge detector. It is self-adapting and is fast, inexpensive, compact and robust.

The CMOS circuit is implemented as a widefield array of 32×32 pixels. At each pixel the received light is compared with an adjacent pixel in order to determine the local light gradient. The result of this comparison is stored and used to connect each pixel to the positive or negative gradient output as appropriate (similar to a balanced knife edge detector). The perturbation of the surface due to ultrasound preserves the speckle distribution whilst deflecting it. The spatial disturbance of the speckle pattern due to the ultrasound is detected by considering each pair of pixels as a knife edge detector. The sensor can adapt itself to match the received optical speckle pattern in less than $0.1\,\mu s$, and then detect the ultrasound within $0.5\,\mu s$ of adaptation. This makes it possible to repeatedly detect ultrasound from optically rough surfaces very quickly. The detector is capable of independent operation controlled by a local microcontroller, or it may be connected to a computer for more sophisticated configuration and control.

We present the theory of its operation and discuss results validating the concept and operation of the device. We also present preliminary results from an improved design which grants a higher bandwidth, allowing for optical detection of higher frequency ultrasound.

1. Introduction

Laser ultrasonics presents an excellent methodology for inspection of industrial components and characterisation of materials because it is non-contact, fast and may be used in hostile environments. The typical setup for all-optical inspection is to use a high energy light pulse to generate an acoustic wave. This acoustic wave propagates as an ultrasound wave, and may be detected using a coherent laser beam. The main parameters of interest for optical detection of ultrasound are achievable bandwidth, sensitivity, dynamic range, linearity of response, spatial resolution, mechanical stability and the surface finish required. A critical parameter of sensitivity is the minimum detectable surface displacement called the noise equivalent surface displacement (NESD). It is possible to achieve sensitivities of the order of 10^{-6}Å$(\text{W/Hz})^{-1/2}$ [1].

Optical detection of ultrasound may be categorised into interferometric or non-interferometric methods [2]. The non-interferometric methods include the optical beam deflection (OBD) technique [3], surface grating technique [4] and reflectivity [5]. Interferometric methods include the use of optical heterodyning, differential interferometry or velocity interferometry [2]. The confocal Fabry-Perot interferometer (CFPi) [6, 7] and adaptive interferometers provide high sensitivity even in the presence of speckle [8, 9].

Recent work in the development of optical detectors has focussed primarily on interferometric techniques in particular for multi-point detection, high adaption rate and industrial applications [10]. The use of two-wave mixing (TWM) in photorefractive crystals (PRC) is well researched [11]. The PRC requires about 2 ms to adapt to the speckle pattern received, making the detectors quite slow, but this technique is able to work effectively with rough surfaces and can work at reasonably high acoustic frequencies (in the region of 1 GHz). The use of photodetector arrays with multiplexed laser interferometers [12, 13] and based on random quadrature demodulation for multi-speckle processing has also been investigated [14]. The interferometric techniques are usually quite vulnerable to low frequency vibrations, expensive, bulky and tend to be quite difficult to set up and keep aligned. Both OBD and interferometric systems work very well when the surface finish is smooth (mirror-like), and offer similar sensitivities. However the OBD technique fails when the surface of the specimen under test is rough [15].

The knife-edge detector (KED) is the simplest form of the OBD detector. It is attractive because it satisfies most of the parameters of interest - particularly achievable bandwidth, linearity of response, mechanical stability, low-cost, small size, and small number of optical components. The KED is most simply implemented with a knife-edge to obscure half of a reflected laser beam that is collected on a photodiode. Use of a balanced or split photodiode provides a differential signal and recent work includes the use of split fibre optic bundles for light collection [16].

To overcome the requirement for a smooth surface for OBD detection, Clark [17] uses a spatial light modulator (SLM) to implement a reconfigurable KED. The probe beam is reflected off the specimen under test onto the SLM. An image of the speckle pattern received on the SLM is captured by a camera, and this image is used to compute a speckle correlated spatial filter - 'an ensemble of knife-edges'. This filter is implemented as a binary mask on the SLM. The mask spatially filters the speckled probe beam and reflects it onto a photodiode for detection of the surface displacement. This system is limited by the frame rate of the camera and SLM to approximately 12 Hz.

In operation, the KED detects the deflection of the probe beam caused by the surface perturbation. The knife edge acts as a boundary or spatial filter, and the detector measures the change in light energy as the beam is deflected across it. It is most commonly implemented as a split photodiode (effectively two photodiodes in very close proximity to one another), and by connecting each of the outputs (labelled L and R) to one input of a difference amplifier, a balanced signal sensitive only to the deflection is obtained. This is shown in figure 1.

2. Optical detection on rough surfaces

Since light is used for detection (the probe beam), the use of lasers for detecting ultrasound is however usually restricted by the surface finish of the material or component under inspection. All surfaces are rough at the atomic level, but of interest to inspection are surfaces which have a roughness of the order of the wavelength of the probe beam used in inspection. The Rayleigh criterion is often used as a criterion for the degree of roughness of a surface [18]. The criterion may be summed up as follows: if $\Delta h < \lambda/8$ the surface is smooth, otherwise it is rough, where Δh is a measure of surface height fluctuation, and λ is the wavelength of the light. Optical smoothness or roughness is dependent on light wavelength and angle of illumination: both factors affect how smooth the surface appears to be.

Thus the optically rough surface presents three problems: firstly, if the surface roughness is comparable with the wavelength of the acoustic signal, then the Rayleigh wave is usually highly attenuated. Secondly, the reflected light is diffusely reflected. Matching the étendue (or

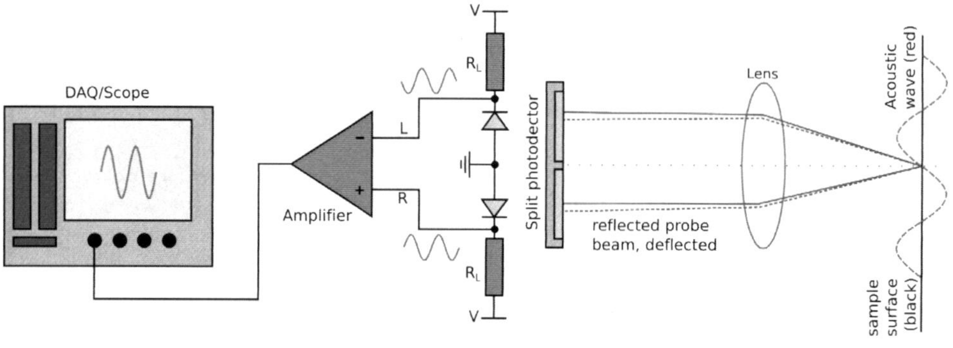

Figure 1. A schematic of a system that uses a knife edge detector (KED) to detect optical beam deflection due to a surface acoustic wave.

throughput - the product of the effective entrance aperture area by the solid angle limited by the rays of maximum inclination) of the detection system to that of the surface space quickly becomes difficult to achieve experimentally [2]. This means that the change in light caused by the surface displacement is usually very difficult to detect because insufficient light is collected.

The third problem that manifests when working with optically rough surfaces is that the variations in the surface under the probe beam reflect the light back with a random phase profile. When this reflected light is captured on a screen or detector, it causes an interference pattern: a speckle pattern. The speckle pattern is a function of the sample surface and the optics. This interferogram causes problems for the KED because the deflection in the reflected probe beam due to the acoustic wave is such that dark and bright parts of the speckle move synchronously, i.e. at the point being inspected, the whole speckle image received is translated in tandem with the acoustic wave. Moving to a different position on the sample results in a new speckle image.

A model illustrates the problem. Figure 2 shows the different scenarios possible. Figure 2(a) shows the light received at the detector when a probe beam with a Gaussian intensity profile has been reflected by a smooth sample surface. The signal sent to each output is shown in a different colour: red is L, and green is R, using the labels in Figure 1. As an acoustic wave passes under the spot illuminated by the probe beam, the wave is deflected, leading to the result shown on the right in Figure 2(b). If a rough surface is inspected, a speckle pattern is received at the detector. Figure 2(c) shows a speckle pattern incident on a KED after reflection off a rough surface. The perturbation due to an acoustic signal results in the acoustic signal shown in Figure 2(d). The signal has a much smaller amplitude compared to Figure 2(b) because both bright and dark parts of the speckle have synchronously crossed the knife edge, resulting in a smaller energy transfer across the knife edge boundary.

Consider a device able to treat each speckle as a separate beam, and thence dynamically able to create a number of knife edge detectors in response to any received speckle pattern. The outputs of these KEDs would be tied together: all Ls together, all Rs together. We call this configuration of linked multiple knife edge detectors the configuration state, as depicted in Figure 2(e). Figure 2(f) shows that the amplitude of the received signal is greatly improved as compared to the KED.

The ideal reconfigurable KED would therefore be a very welcome device, allowing non-contact inspection of samples without the need for extra processing to enable the sample surface to satisfy the Rayleigh criterion. Conceptually, the detector would analyse the gradient of the received light and create splits at the points of maximum intensity. It could do this in both vertical or horizontal axes—corresponding to the orientation of a regular KED. The configurations shown

Journal of Physics: Conference Series **581** (2015) 012009 doi:10.1088/1742-6596/581/1/012009

in figure 2 are for a horizontal axis.

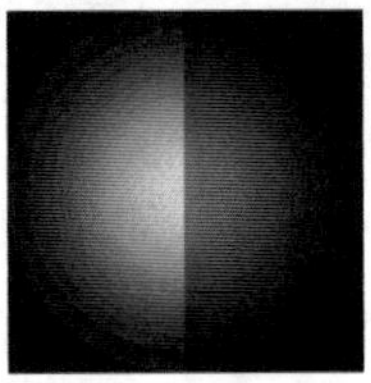

(a) Light from smooth surface received on KED

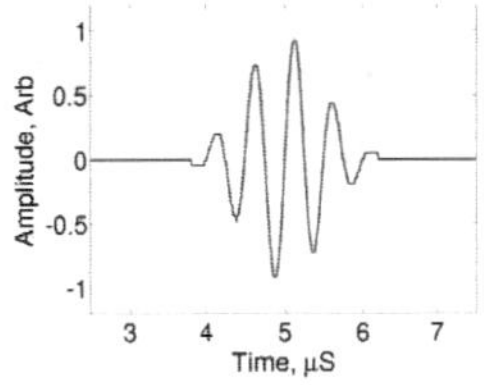

(b) Detected signal

(c) Light from rough surface received on KED

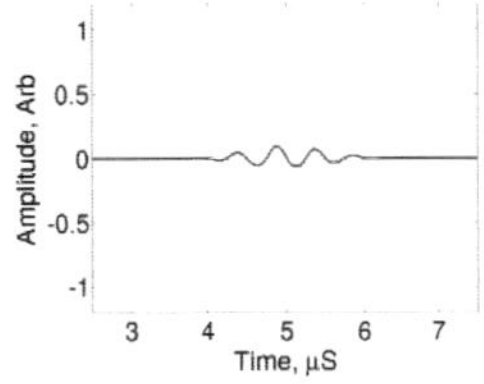

(d) Detected signal

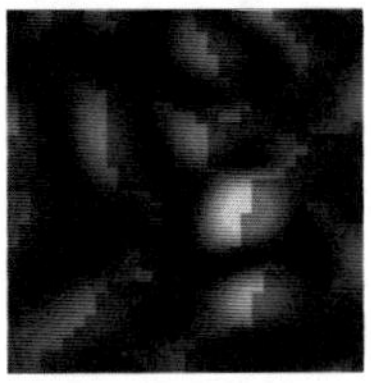

(e) Light from rough surface received on configurable KED

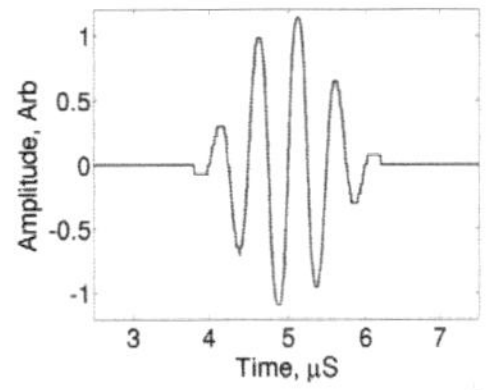

(f) Detected signal

Figure 2. Light received at a KED and detected signal for configurations of surface roughness. (a) shows light from smooth sample inspected with a KED and the result obtained (b), (c) is for a rough sample inspected with a KED and result obtained (d), and (e) is shows speckle from a rough sample inspected with a reconfigurable KED, and the result obtained (f). The total light energy is the same for each of the 3 situations.

3. The SKED - Speckle Knife Edge Detector

The Speckle Knife Edge Detector (SKED) reported here is a detector that implements the automatic self configuration process outlined above. It is implemented as an integrated circuit and is fabricated in a 0.35 μm standard CMOS process. It is laid out as an array of 32×32 smart pixels. Each pixel is 59 μm×59 μm, 43 % of which is photosensitive. Each pixel contains an N-well photodiode (the light-sensitive region), signal conditioning, a comparator, and 1 bit memory (for storing configuration states). It also contains 3 switches implemented as transmission gates. This amounts to 78 transistors per pixel. A simplified schematic of the pixel is shown in figure 3.

The SKED has two modes of operation. In *configure* mode, each pixel of the SKED compares the light received to the light received by the pixel adjacent to it. The adjacent pixel to it is defined by the axis of comparison (vertical or horizontal), which is selectable by the user. The comparator output is used to determine which of the L or R outputs (equivalent to red or green in figure 2) that the photodiode for this pixel should be connected to. This result is stored in

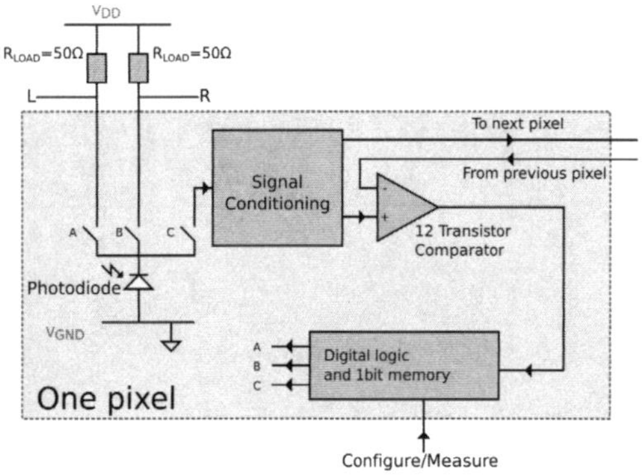

Figure 3. Schematic of a pixel of the SKED.

Figure 4. SKED (in red outline) on a printed circuit board.

the memory at this pixel. It is possible to send a signal to each pixel to explicitly define which output the photodiode should be connected to. The L outputs from all the pixels, and R outputs from all the pixels are tied together and presented as two outputs outside the chip. This gives the sum of all positive gradients in L, and sum of all negative gradients in R, equivalent to a normal KED. In the *measurement* mode of operation, the comparator is disconnected, and the photocurrent is switched through to the appropriate output at each pixel.

Figure 4 shows the printed circuit board laid on an optical table. The SKED chip is outlined in red. Figure 5 shows an outline of the decision making process. It shows 7 pixels, each receiving different light intensities. Since there are two peaks in the light intensity distribution, each of those peaks will act as a split point for a knife edge. The decisions taken by the SKED are shown in the same figure.

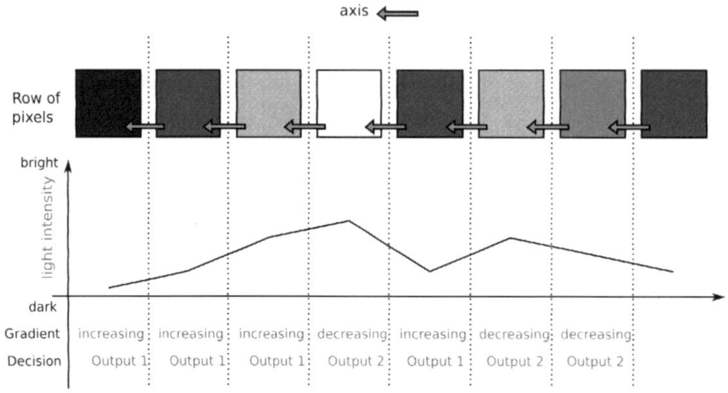

Figure 5. A schematic of operation of the SKED in *configure* mode.

The SKED is small, very fast, rugged and easy to swap into existing systems which use a KED. It is relatively inexpensive, with 9500 Euro providing 30 fabricated devices. It is hosted on a credit-card sized printed circuit board with microcontroller providing control and USB connectivity. It is possible to read the configuration state of each pixel (i.e. which output the pixel is connected to - L or R), or to write arbitrary patterns to the SKED, such as a KED with a diagonal axis.

In normal operation, at the start of the scan the SKED configures itself to the received speckle pattern, then goes into measurement mode where it is ready to measure the acoustic signal. The sample may then be moved, at which point the speckle pattern changes completely, and it becomes necessary to reconfigure to this pattern before the next acoustic signal is measured. A raster scan may then be performed, alternating between *configure* and *measurement* modes at each point.

4. Results

This section presents some prelimnary results obtained with the SKED. A commercially available detector from Bossa Nova, the Tempo [19], was used as a reference device to calibrate the SKED.

4.1. Adaption Speed

The SKED can configure itself to the received speckle in as little as 0.1 μs, and be ready to measure ultrasound within 0.5 μs of configuration. This means that it is possible to take as many as 1,000,000 measurements per second. This makes it extremely fast compared to existing technology such as the Tempo. The Tempo needs approximately 2 ms to reconfigure to speckle.

4.2. Sensitivity

The sensitivity of the SKED is affected by how each pixel correctly configures to the light received, connecting the photodiode to the appropriate output (L/R). It has been determined that the SKED does this correctly approximately 90% of the time, under typical light power conditions, and will correctly choose which output when there is as little as 18 nW difference in light intensity between two adjacent pixels.

4.3. Bandwidth

The bandwidth of the device is limited by the capacitance of the photodiode and the parasitic impedances from the switches and tracks. It was determined that when balanced, i.e. with 512 pixels connected to each output (L/R), the -3 dB bandwidth is approximately 16 MHz. We have successfully measured signals at 82 MHz, but observed attenuation due to the bandwidth.

4.4. Comparison with a KED

Since it is possible to write configuration patterns to the device, it is possible to configure it to operate as a regular KED. This allows the performance of the SKED to be compared with a KED without other factors affecting the results (such as total photosensitive area). Figure 6 shows the improvement that can be achieved when the SKED operates in the self-configured mode. An 82 MHz surface acoustic wave was generated on an aluminium sample with a machined finish. A 1 mm line scan was performed across the surface, and at each point the signal was measured both with the SKED configuring itself to the received speckle pattern, and programmed as a KED. The mean signal and amplitude show a clear improvement over the KED operation.

4.5. Improved design - SKED2

Following from the lessons of the prototype device, an improved second device was designed and fabricated. It was fabricated in the same 0.35 μm CMOS process. Each pixel is 62 μm$\times$62 μm, with a photosensitive area 27% of the pixel area. It also has 32$\times$32 pixels. The main improvements are as follows:

(i) Simplified comparator: the comparator accounted for a significant area of the prototype device. It was simplified from 12 to 4 transistors. This improved both the space utilisation and improved the adaption speed.

(ii) Reduced parasitics: the parasitic impedance in the prototype device were quite high, and by simplifying the switches and reducing the track width this was reduced.

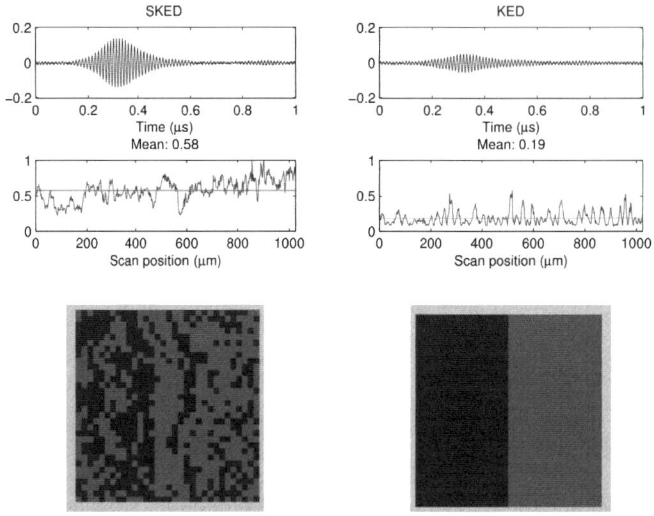

Figure 6. On the left, the SKED device automatically configures itself as a line scan is performed on machined aluminium; on the right, the SKED device is manually configured to behave as a conventional KED. The upper traces are the ultrasonic signals at the last point in the line scan. The centre traces show the mean signal level, normalised to the maximum of all recorded signals. The images at the bottom represent the configuration state for the last point in the line scan.

(iii) Improved bandwidth: a more radical change was the introduction of a second 4-transistor comparator in each pixel. This comparator is used to compare the photocurrent with a reference current supplied from the printed circuit board. The reference can be changed to match the experiment. In *configure* mode, this comparator is used to determine whether the light intensity at the pixel exceeds the reference value. The output of the comparator determines whether this pixel is connected or disconnected to the outputs in *measurement* mode. This is useful because if a photodiode does not receive much light (i.e. dark speckle), it does not contribute signal, but the capacitance of the photodiode reduces the bandwidth of the device. By disconnecting pixels in this way, the bandwidth of the device can be improved without sacrificing signal.

Figure 7 shows an experimental result that demonstrates the value of being able to disconnect dark pixels from the chip outputs. The reference current applied to the chip is varied whilst measuring a single sample point, hence a constant speckle pattern. The black region indicates pixels which receive low light and have been disconnected from the outputs. At both 10 MHz and 20 MHz, as more and more pixels are disconnected by increasing the reference current (the SKED chooses which pixels to switch off), the received signal amplitude increases, until the reference level is high enough to start disconnecting pixels that have significantly high illumination levels and not enough light is detected. At this point the amplitude begins to fall.

This new design is also more sensitive to differences in the received light between adjacent pixels, with correct configuration when there is as little as 6 nW difference in light power.

5. Conclusion

We have presented a CMOS integrated circuit capable of optical detection of ultrasound from optically rough surfaces. It is robust, cheap and adapts very quickly. Preliminary results

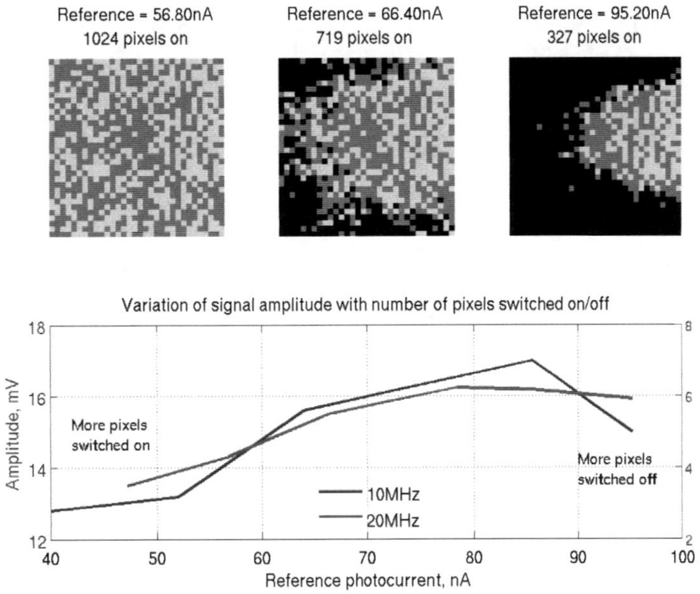

Figure 7. Improvement in detected signal by switching off pixels.

demonstrate the improvement in detection capability as compared with conventional detection techniques based on optical beam deflection. Initial experiments indicate a noise equivalent surface displacement of 12×10^{-6}Å$(W/Hz)^{1/2}$ for the SKED compared with 2×10^{-6}Å$(W/Hz)^{1/2}$ for the Tempo. Full specifications will be published soon.

6. Acknowledgements
The authors acknowledge the support of the Research Centre in Non-Destructive Evaluation UK, for this work.

References
[1] Sontag H and Tam A C 1986 Optical detection of nanosecond acoustic pulses *IEEE Trans.Ultrasonics, Ferroelectrics and Frequency Control* **33** 500–6
[2] Monchalin J P 1986 Optical detection of ultrasound *IEEE Transactions on Ultrasonics, Ferroelectrics and Frequency Control* **33** 485–99
[3] Adler R, Korpel A and Desmares P 1968 An instrument for making surface waves visible *IEEE Trans on Sonics and Ultrasonics* **15** 157–60
[4] Korpel A, Laub L J and Sievering H C 1967 Measurement of acoustic surface wave propagation characteristics by reflected light *Appl. Phys. Lett.* **10** 295–7
[5] Rosencwaig A, Opsal K, Smith W L and Willenborg D L 1985 *Appl. Phys. Lett.* **46** 1013–5
[6] Monchalin J P and Heon R 1989 Broadband optical detection of ultrasound by optical sideband stripping with a confocal Fabry-Perot *Appl. Phys. Lett.* **55** 1612–4
[7] Blouin, A., Padioleau C, Nron C, Lvesque D and Monchalin J P 2007 Diffferential confocal fabry-perot for the optical detection of ultrasound *AIP Conf. Proc* **894** 193–200
[8] Dewhurst R J and Shan Q 1999 Optical remote measurement of ultrasound *Meas. Sci. Tech.* **10** R139–68
[9] Nakano H, Matsuda Y, Shin S and Nagai S 1995 Optical detection of ultrasound on rough surfaces by a phase-conjugate method *Ultrasonics* **33** 261–4
[10] Pouet B, Wartelle A and Breugnot S 2011 Multi-detector receiver for laser ultrasonic measurement on the run *Nondestructive testing and evaluation* **26** 253–66
[11] Kamshilin A A 2002 Adaptive photorefractive interferometers for ultrasound detection *Proc. SPIE* **4919** 29–39

[12] Murray T W and Krishnaswamy S 2001 Multiplexed interferometer for ultrasonic imaging applications *Opt. Eng* **40** 1321–8
[13] Zhou Y, Murray T W and Krishnaswamy S 2002 Photo-acoustic imaging of surface acoustic wave slowness using multiplexed, two-wave mixing array interferometry *IEEE Trans. Ultrasonics, Ferroelectrics and Frequency Control* **49** 1118–23
[14] Pouet B, Breugnot S and Clemenceau P 2006 Robust laser ultrasonic interferomter based on random quadrature demodulation *AIP Conf Proc* **820** 233–9
[15] Whitman R L and Korpel A 1969 Probing of acoustic surface pertubations by coherent light *Applied Optics* **8** 1567–76
[16] Murfin A S, Soden R A J, Hatrick D and Dewhurst R J 2000 Laser-ultrasound detection systems: a comparative study with Rayleigh waves *Meas. Sci. and Tech.* **11** 1208-1219
[17] Clark, M 2011 Optical detection of ultrasound on rough surfaces using speckle correlated spatial filtering *Journal of Physics: Conf Series* **278** 012025
[18] Rayleigh L 1877 The Theory of Sound *New York:* **Macmillan**
[19] Wartelle A, Pouet B and Breugnot, S 2011 Non-destructive testing using two-component/two-wave mixing interferometer *AIP Conf. Proc. 1335* **47** 265–72

Fabrication of Cheap Optical Transducers (CHOTs) on film carriers for in-situ application and generation of surface acoustic waves

V Ageeva, T Stratoudaki, M Clark and M G Somekh

Electrical Systems and Optics Division, Faculty of Electrical and Electronic Engineering, University of Nottingham, Nottingham, NG7 2RD, United Kingdom
Phone: +44 115 9515605, Fax: +44 115 9515616;

E-mail: eexva1@nottingham.ac.uk

Abstract. Cheap optical transducers (CHOTs) are patterns on the surface of a component activated by lasers to generate and detect ultrasound. Excited optically, with minimal surface impact, and fully customizable, CHOTs provide a simple alternative to conventional piezoelectric transducers, offering wireless, remote operation. Of particular interest is application of CHOTs for in-situ ultrasonic inspection of hard-to reach and complex-geometry components such as those of aero-engines. A suitable fabrication method has been developed to allow in-situ application of CHOTs onto large size and curved components, as well as those already in service, challenging for current laboratory-based micro-patterning methods. This work describes the fabrication of a transferable g-CHOT for generation of ultrasound. The g-CHOT has been made on an SU8 carrier film using a sacrificial polystyrene layer, allowing the transducer to be transferred from the substrate and subsequently delivered and applied to the surface of the sample in-situ. The functionality of the fabricated transducer is demonstrated by detection of the Surface Acoustic Waves (SAW) generated by the g-CHOT transferred onto glass and aluminium samples.

1. Introduction

Cheap Optical Transducers (CHOTs) are nanometer-height patterns attached to or printed on a component and activated optically using laser light to generate (g-CHOT) and detect (d-CHOT) ultrasound [1]. Using principles of laser ultrasonics [2] they provide a simple alternative to conventional piezoelectric transducers (PZTs) in instances where remote non-contact operation is required. Additionally, used with the endoscopic CHOTs pulser [3], they offer simplified and flexible probe instrumentation when the access path and size considerations restrict the use of other laser ultrasonic techniques [4-6], providing greater efficiency [7]. Demanding applications, such as the in-situ non-destructive testing (NDT) in aerospace industry, frequently require testing of components with surfaces of complex geometry, hard-to-reach components in locations with limited access or in hot, hazardous environments. These conditions are equally challenging for the PZTs due to their contact nature as well as for the spatially-consuming instrumentation of the laser ultrasonic detection methods [6], restricting the use of both techniques.

The endoscopic CHOT system [3] combines the use of CHOTs with the fibre light delivery to enable in-situ ultrasonic testing of the aero-engine components. Its development has been paired with the upgrade of the CHOT manufacturing methods to satisfy industrial requirements for the in-situ application of the transducers. This paper presents the fabrication method and functionality test results of a transferable g-CHOT, produced on a carrier film to enable transducer transfer to components and in-situ application, previously limited to a laboratory environment. Overcoming the restrictions of the direct sample-patterning methods used previously also allows application of CHOTs onto large and curved components, complementing the endoscopic system.

2. CHOTs for in-situ application and testing

The g-CHOT pattern, used for generation of ultrasound, consists of alternating areas of high and low absorption converting incident radiation energy into acoustic waves [1, 8, and 9] via a photo-thermal effect. The g-CHOT design determines the characteristics of the generated ultrasound [8] and the pattern geometry can be configured to produce plane or focused, surface or bulk acoustic waves [9] of desired frequency.

CHOTs could be manufactured by a number of micro-patterning methods. Although such methods (e.g. photolithography, direct material deposition through masks) provide precise feature control and repeatability, they are hard to implement with large size or curved parts due to process complexity or equipment requirements, and they could not be applied to components in-situ. This task could be achieved by alternative methods that offer in-situ fabrication of CHOTs (laser etching or laser induced forward transfer (LIFT)). Another method is the creation of a transferable CHOT, described in this work, which combines the control and repeatability of the laboratory-based fabrication with the flexibility and ease of the in-situ application.

Transferable CHOT is created by producing a CHOT pattern on a carrier film separated from the substrate with a sacrificial layer (figure 1(a)). The sacrificial layer is later dissolved releasing the CHOT carrier (figure 1(b)) and allowing the transducer to be transferred and applied to components when and where it is required.

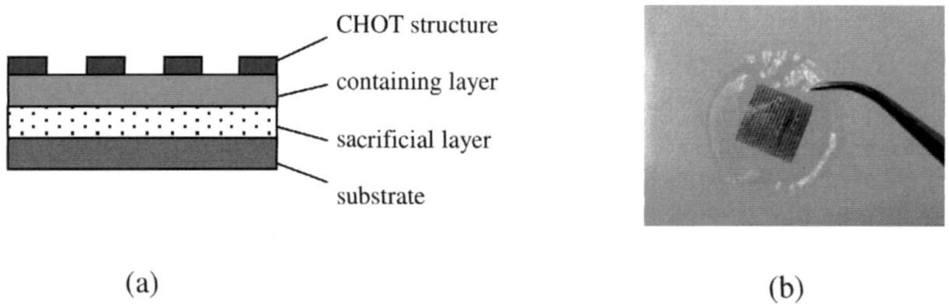

Figure 1. Transferable g-CHOT. (a) Schematic of the fabricated structure, (b) Released g-CHOT on an SU8 carrier.

The use of polystyrene (PS) sacrificial layers is simple, provides fast etching times [10], and has been previously demonstrated [11] for successful release of structures from photoresist SU8. The transferable CHOTs described in this work were produced on SU8 carriers using PS sacrificial layers following similar and adapted procedure. The use of SU8 as a CHOT carrier suits the required application providing film transparency necessary for transducer operation.

In the work presented here, transferable g-CHOTs for generation of 5MHz plane SAWs on aluminium (Al) samples were made of 70nm of Chromium (Cr) with the corresponding spatial period of the pattern ~720μm and the pattern size 10x10mm.

3. Method

Precision glass windows from Comar (Ø25mm) were used as substrates. Polystyrene solution was prepared, spin-coated onto the cleaned substrates and baked to create the sacrificial layer. The containing layer for the transducer was then created by spin-coating of SU8 on top of the glass-PS structure. The resulting structure with SU8 was then pre-baked, exposed to UV and post-baked according to manufacturer's guidelines [12]. The g-CHOT pattern was then produced onto the glass-PS-SU8 structure by direct deposition of 70nm of Chromium in a sputterer through a mask.

The sacrificial layer of polystyrene was finally removed by immersion of the samples in toluene for approximately 5 hours, after which the film containing the transducer was released and dried. Fabrication stages, materials and parameters used are shown in Table 1.

Table 1. Fabrication process of a g-CHOT on an SU8 carrier.

Sacrificial layer PS 100k 10% solution ⇒	**Containing layer** SU8-100 ⇒	**CHOT** Cr ⇒	**Carrier separation** Toluene
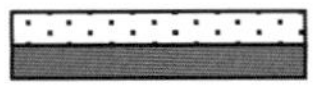	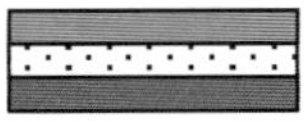	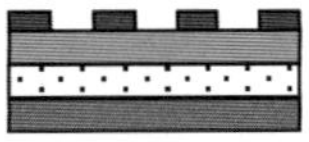	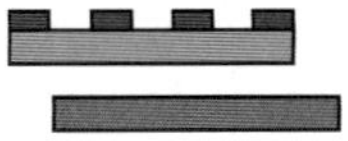
Spin-coat: 30” at 700rpm	Spin-coat: 10” at 500rpm 30” at 2500rpm	Deposit metal in sputterer: 70nm	Immerse structure: 5hrs + agitation
Bake: 1 hr at 100°	Pre-Bake Expose Post-bake Develop		

4. Fabricated transducer

The functionality of the fabricated transducer was demonstrated by detection of the SAWs generated by the transferred g-CHOT under laser illumination (figure 2). The g-CHOT structure, coupled to a glass sample with industrial gel couplant, was activated by illumination from a q-switched Nd:YAG with average power of 450mW at λ=1.064 μm and 1kHz repetition rate. The detection of the generated acoustic waves was performed by a d-CHOT produced directly on the

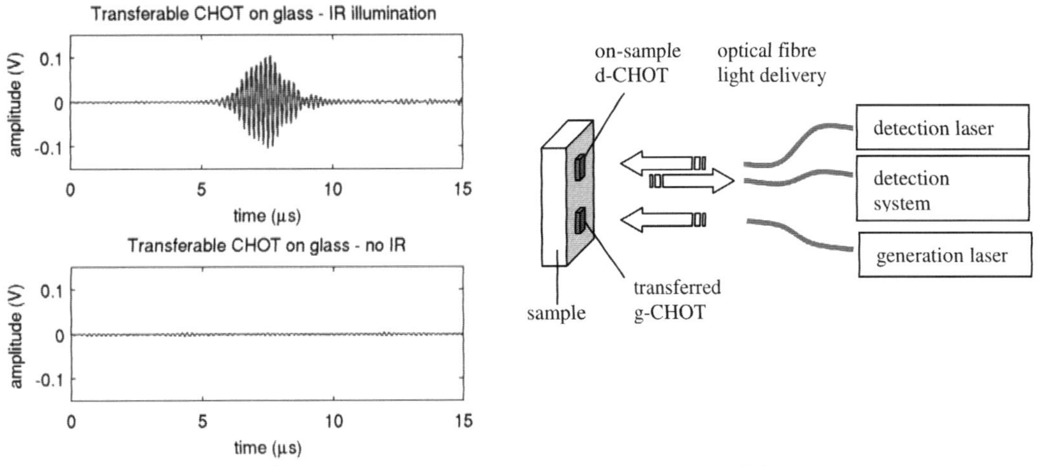

(a) (b)

Figure 2. Functionality test. (a) SAWs detected upon laser illumination of the transferred CHOT (top), and the signal recorded in absence of laser illumination (bottom), (b) Experimental setup.

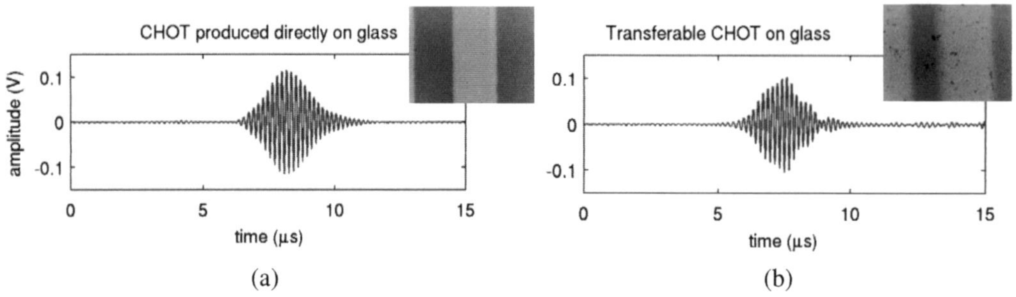

Figure 3. SAWs detected from a g-CHOT produced directly on a sample (a) and from a g-CHOT on a carrier transferred onto the sample (b); microscope images of the corresponding CHOT structures in reflection, with the light areas corresponding to Cr (insets).

sample surface and illuminated by a continuous wave second harmonic Nd:YAG emitting at 0.532μm and providing 120mW power. Signal acquisition was performed using a 5MHz band-pass filter, averaging 1000 waveforms and using 2xZFL500-LN amplifiers from Minicircuits. Figure 3 shows a comparison to the SAWs detected from a g-CHOT produced by direct sample patterning.

The ultrasonic generation by a transferred g-CHOT could result from the ultrasonic source located a) on a carrier, with the g-CHOT absorbing the radiation, and b) on the sample surface with the g-CHOT spatially modulating laser illumination and absorption by the sample [13]. The detection of SAWs shown on the glass samples confirms the g-CHOT as the ultrasonic source due to the absence of absorption by glass.

Further investigation on aluminium samples allows isolation of the two generation mechanisms by reversing the sides with which the transducer is coupled to the sample, and therefore changing the role of the g-CHOT from a transducer to a mask (figure 4). The clear detection of SAWs in the first case (g-CHOT=transducer) and absence in the second (g-CHOT=mask) confirms the preservation of the generation mechanism employed by the g-CHOTs produced by direct sample patterning. A Polytec vibrometer (OVF534) and a 5MHz band-pass filter were used for detection, with the generation setup as described previously.

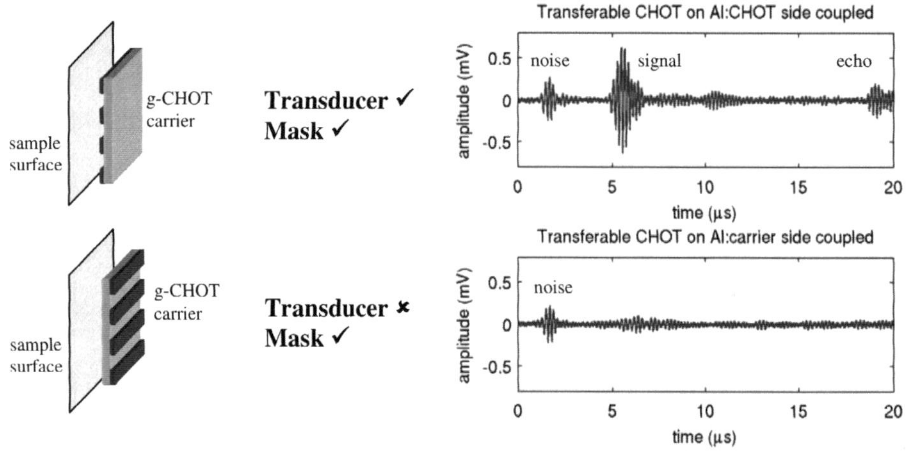

Figure 4. Generation mechanisms demonstrated on Al samples. Top: signal recorded with the patterned surface coupled to the sample, Bottom: signal recorded with the carrier surface coupled to the sample.

5. Conclusions and further work

This work presented a fabrication method developed to produce a transferable g-CHOT, Cheap Optical Transducer used for generation of ultrasound, which could be applied in-situ. The developed method allows the sensor to be fabricated on a carrier film in controlled laboratory environment and consequently be transferred and applied to components where and when it is required. This enables CHOTs to be used for testing of large, curved-surface components, as well as providing the means of sensor delivery to the parts already in service, and complementing previously developed endoscopic CHOTs pulser. Functionality of the fabricated transducer was shown on glass and Al samples, with comparative results from the g-CHOTs produced on samples directly.

Increasing the flexibility of the CHOT carrier and developing more permanent means of transducer attachment is currently under investigation, as well as the fabrication of a transferable d-CHOT for detection of SAWs.

Acknowledgements

The authors wish to thank the Engineering and Physical Sciences Research Council and Rolls-Royce plc for their support.

References

[1] Stratoudaki T, Hernandez J, Clark M and Somekh M G 2007 Cheap optical transducers (CHOTs) for narrowband ultrasonic applications *Meas. Sci. Technol.* **18** pp 843-851

[2] Scruby C B and Drain L E 1990 Laser ultrasonics, techniques and applications *Adam Hilger* Bristol UK

[3] Ageeva V, Stratoudaki T, Clark M and Somekh M G 2013 Integrative solution for in-situ ultrasonic inspection of aero-engine blades using endoscopic cheap optical transducers (CHOTs) *The e-Journal of Nondestructive Testing* ISSN 1435-4934 **18** No 12

[4] Davies S J, Edwards C E, Taylor G S and Palmer S B 1993 Laser generated ultrasound: its properties, mechanisms and multifarious applications *J. Phys. D: Applied Physics* **26** pp329-348

[5] Dewhurst R J and Shan Q 1999 Optical remote measurement of ultrasound *Meas. Sci. Technol* **10** pp 139-168

[6] Monchalin J P 1986 Optical detection of ultrasound *IEEE Trans. Ultrason. Ferroelectr. Freq. Control* **33** pp 485-499

[7] Murray T W, Deaton J B and Wagner J W 1996 Experimental evaluation of enhanced generation of ultrasonic waves using an array of laser sources *Ultrasonics* **34** pp 69-77

[8] Arca A, Aylott J, Marques L, Clark M, Somekh M G, Smith R, Sharples S, Stratoudaki T and Chen X 2011 CHOTs optical transducers *Nondestructive Testing and Evaluation* **26** No 3-4 pp 353-366

[9] Stratoudaki T, Arca A, Clark M and Somekh M G 2012 Cheap optical transducers (CHOTs) for generation and detection of longitudinal waves *IEEE Proceedings* 10 D.O.I 10.1109 pp 961-964

[10] Psoma S D and Jenkins D W K 2005 Comparative assessment of different sacrificial materials for releasing SU8 structures *Rev.Adv.Mater.Sci.* 10 pp 149-155

[11] Cheng L, Govindaraju A, Garra J, Schneider T, White R, Currie J and Paranjape M 2004, Releasing SU-8 structures using polystyrene as a sacrificial material, *J.Sna* A 114 Elsevier pp 123-128

[12] MicroChem SU-8 50-100 Data Sheet Rev 2.02, available: http//www.microchem.com/Prod-SU82000.htm [accessed 24th March 2014]

[13] Wagner J W, Mckie A D W, Spicer J B, Deaton Jr J B 1990 Modulated laser array sources for generation of narrowband and directed ultrasound *Journal of Nondestructive Evaluation* **9** No 4 pp 263-270

Guided wave propagation and scattering in pipeworks comprising elbows: Theoretical and experimental results

M El Bakkali[1], A Lhémery[1], V Baronian[1] and B Chapuis[1]

CEA, LIST – F- 91191 Gif-sur-Yvette, France

E-mail : alain.lhemery@cea.fr

Abstract. Elastic guided waves (GW) are used to inspect pipeworks in various industries. Modelling tools for simulating GW inspection are necessary to understand complex scattering phenomena occurring at specific features (welds, elbows, junctions…). In pipeworks, straight pipes coexist with elbows. GW propagation in the former cases is well-known, but is less documented in the latter. Their scattering at junction of straight and curved pipes constitutes a complex phenomenon. When a curved part is joined to two straight parts, these phenomena couple and give rise to even more complex wave structures. In a previous work, the Semi-Analytic Finite Element method extended to curvilinear coordinates was used to handle GW propagation in elbows, combined with a mode matching method to predict their scattering at the junction with a straight pipe. Here, a pipework comprising an arbitrary number of elbows of finite length and of different curvature linking straight pipes is considered. A modal scattering matrix is built by cascading local scattering and propagation matrices. The overall formulation only requires meshing the pipe section to compute both the modal solutions and the integrals resulting from the mode-matching method for computing local scattering matrices. Numerical predictions using this approach are studied and compared to experiments.

1. Introduction

Thanks to their ability to propagate over long distance without attenuation, elastic guided waves (GW) are used in efficient non-destructive testing (NDT) operations for the in-service inspection of industrial pipeworks. However, because of the multi-modal and dispersive properties of GW, scattering phenomena involved when GW interact with specific features in the pipework are complex. Tools for simulating non-destructive examinations are thus useful to understand measured signals and to design or improve testing configurations.

In a straight pipe, GW propagation is well known: efficient computational methods are used to predict them such as the Semi-Analytic Finite Element method (SAFE) or the Finite Element (FE) method. Conversely, the GW propagation in curved guides, which are common in industrial plants, is less documented [1-2] and more difficult to model. In pipeworks, curved guides coexist with straight ones; hence, the scattering of GW at their junction is an important problem.

The aim of the present work is to study numerically and experimentally the GW propagation in a pipe elbow linked to straight parts using a modal approach in order to supplement capabilities of CIVA platform [3]. For this, a numerical model able to deal with curved guides is developed. This model is based on an extension of the SAFE method, commonly used to simulate the GW propagation in straight guides. Once the modal solution for GW in an elbow is obtained, the scattering at the junction between a straight and a curved guide is investigated. A numerical model based on the mode matching method is used to predict the modal decomposition of an incident mode scattered at such junction. Then, a global scattering matrix is built, accounting for the scattering at several junctions and for the guided propagation in between. An example of parametric study is given to illustrate the model capabilities. Experiments on an elbow of typical industrial characteristics are made for validation.

Mode conversions arising by the scattering of guided waves at the junction between a curved guide and a straight guide have been investigated. Finally, numerical results are compared to experimental studies in order to illustrate the capabilities of this model.

2. Semi-analytical model for guided waves propagations in pipe works comprising elbows

In this section the overall modelling approach is summarized. Readers interested in its detailed derivation are referred to Ref. 4.

2.1. SAFE model in curvilinear coordinate system

The SAFE method is a numerical model used to describe the GW propagation in a waveguide in terms of a modal decomposition of elastodynamic fields. The GW modal basis is the solution of the wave equation in an elastic guide assumed to be homogeneous and to have free boundaries. The equation of motion writes:

$$\begin{cases} -\mathbf{div}\sigma(\mathbf{u}) - \omega^2 \rho \mathbf{u} = 0 & \text{in } \Omega, \\ \sigma(\mathbf{u}) \cdot \mathbf{n} = 0 & \text{on } \Sigma_1, \end{cases} \quad . \tag{1}$$

where $\mathbf{u}$ is the displacement, ρ the mass density, ω the pulsation, Ω the guide section and Σ_1 its guiding surface. At a given frequency, the semi-analytical finite element (SAFE) method allows solving Eq. (1) in the cross section of the guide by a finite element computation, the dependency upon the guiding axis being handled analytically, assuming invariant cross section and material properties along the propagation axis.

Contrary to the case of a straight guide, the Cartesian basis is inadequate to satisfy this condition for a curved guide. Hence, the SAFE formulation has to be written in a curvilinear coordinates system, also known as the Serret-Frenet basis. Figure 1 illustrates the new coordinate system for a curved waveguide (a ring torus):

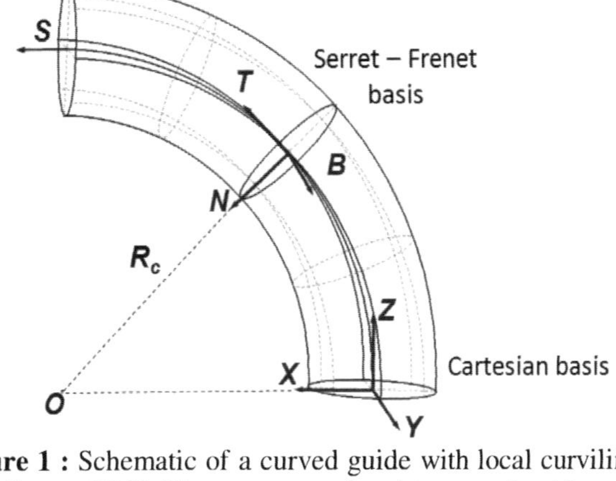

Figure 1 : Schematic of a curved guide with local curvilinear coordinates (N, B, T) system associated to curved guide.

The three axes of this local basis are obtained from the expression of the position vector defining the centerline of the curved guide. The metric tensor of this covariant basis, necessary to define the elementary volume, is given by:

$$\mathbf{g} = \begin{pmatrix} 1 & 0 & 0 \\ 0 & 1 & 0 \\ 0 & 0 & (1-\gamma x)^2 \end{pmatrix} \tag{2}$$

where $\gamma = 1 / R_c$ is the curvature and x the coordinate in the normal direction $\mathbf{N}$ of each node of the guide section.

To integrate the variational formulation in a curvilinear basis, an elementary volume corresponding to the coordinate differentials (*dx*, *dy*, *ds*) must be expressed in this basis [5]. This volume *dV* is defined using $\mathbf{g}_1$, $\mathbf{g}_2$ and $\mathbf{g}_3$ the axes of the non-orthonormal covariant basis associated to the curvilinear basis as follows:

$$dV = \left|\mathbf{g}_1 .(\mathbf{g}_2 \wedge \mathbf{g}_3)\right| dxdyds = \sqrt{\det(\mathbf{g})}\, dxdyds \qquad (3)$$
$$\text{with } \mathbf{g}_1 = \mathbf{N}, \mathbf{g}_2 = \mathbf{B} \text{ and } \mathbf{g}_3 = (1-\gamma x)\mathbf{T}$$

Assuming a time harmonic $e^{(-j\omega t)}$ dependency and similarly to the classical case of a straight guide in the Cartesian coordinates system, the 3D variational formulation must first be written in the non-orthonormal covariant basis. Then, using the relation between the covariant and the curvilinear basis, each term of the variational formulation can be expressed in the curvilinear (**N**, **B**, **T**) basis. The final expression is given by:

$$\int_\Omega \delta\boldsymbol{\varepsilon}_{\mathbf{cur}}^{\mathbf{T}} \mathbf{C} \boldsymbol{\sigma}_{\mathbf{cov}} \sqrt{\det(\mathbf{g})}\, dxdyds - \omega^2 \int_\Omega \rho\, \delta\mathbf{u}_{\mathbf{cur}}^{\mathbf{T}} \mathbf{u}_{\mathbf{cur}} \sqrt{\det(\mathbf{g})}\, dxdyds = 0, \qquad (4)$$
$$\text{with } \mathbf{u}_{\mathbf{cur}}(x,y,s) = \mathbf{u}_{\mathbf{cur}}(x,y)e^{i(ks-\omega t)} \quad \text{and} \quad \delta\mathbf{u}_{\mathbf{cur}}(x,y,s) = \delta\mathbf{u}_{\mathbf{cur}}(x,y)e^{-i(ks-\omega t)},$$

with ε_{cur}, σ_{cur} and u_{cur}, are the strain, stress and displacement fields in curvilinear coordinates.

Once the variational formulation (4) has been written in the curvilinear basis, the above-mentioned condition is satisfied and the *s* dependency of the displacement is restricted to the argument of the exponential of propagation, $e^{(jks)}$, which can be factorized in all elastodynamic field components. From this, the 3D variational formulation can be turned into a semi-analytical variational formulation by restricting the volume integration to a surface integration over the guide cross section. Finally, the numerical expression is written as:

$$(\mathbf{K}_1^e + j\beta\mathbf{K}_2^e + \beta^2\mathbf{K}_3^e)\mathbf{U} - \omega^2\mathbf{M}^e\mathbf{U} = 0, \qquad (5)$$

where **U** is the displacement vector and β the wavenumber. $\mathbf{K}_{i(i=1,2,3)}^e$ and $\mathbf{M}^e$ are the elementary stiffness and mass matrices in the (**N**, **B**, **T**) basis, respectively, expressed as:

$$\mathbf{K}_1^e = \int_{S_e} \mathbf{N}^{e\mathbf{T}}\mathbf{L}_{xy}^{\mathbf{T}}\mathbf{C}\mathbf{L}_{xy}\mathbf{N}^e \sqrt{\det(\mathbf{g})}dxdy, \quad \mathbf{K}_2^e = \int_{S_e} \mathbf{N}^{e\mathbf{T}}\mathbf{L}_{xy}^{\mathbf{T}}\mathbf{C}\mathbf{L}_{s}\mathbf{N}^e \sqrt{\det(\mathbf{g})}dxdy,$$
$$\mathbf{K}_3^e = \int_{S_e} \mathbf{N}^{e\mathbf{T}}\mathbf{L}_{xy}^{\mathbf{T}}\mathbf{C}\mathbf{L}_{s}\mathbf{N}^e \sqrt{\det(\mathbf{g})}dxdy, \quad \mathbf{M}^e = \int_{S_e} \rho\mathbf{N}^{e\mathbf{T}}\mathbf{N}^e \sqrt{\det(\mathbf{g})}dxdy. \qquad (6)$$

$\mathbf{N}^e$ is the finite element interpolation matrix and $\mathbf{L}_{xy}^{\mathbf{T}}$, $\mathbf{L}_{s}^{\mathbf{T}}$ are the elementary matrices introducing the curvature effect defined by Treyssède and Laguerre [6]. For a zero-curvature, these matrices exactly equal the well-established matrices for a straight waveguide.

At a given frequency and both in a straight and in a curved guide, the solution of the quadratic eigenvalue system of Eq. (5) allows us to compute the modal solution by calculating the eigenvalues and the eigenvectors, which correspond respectively to the wavenumbers and the vectors of particle displacement of every mode (propagative, evanescent and inhomogeneous).

2.2. The mode matching model at the junction of two guides

In a pipework comprising curved parts, pipe elbows are generally joined to straight pipes. When guided waves propagate, scattering phenomena arise at such junctions. Here, we aim at computing the corresponding scattering matrix for predicting the reflection and transmission phenomena occurring at the junction as illustrated by figure 2.

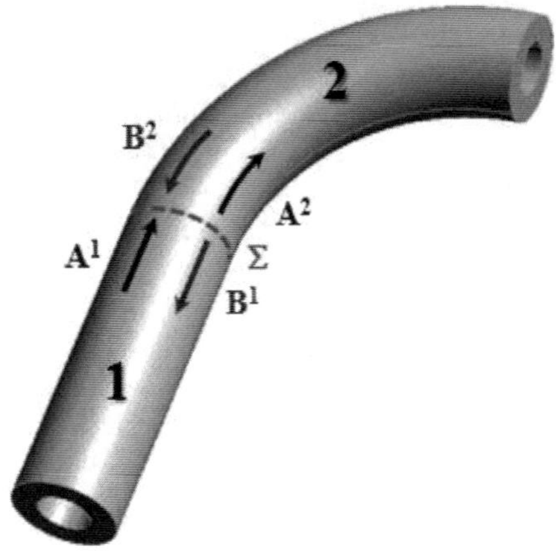

Figure 2 : Incoming and outgoing families of modes at the junction between the straight and the curved guides.

Assuming the completeness of the calculated eigenmodes obtained using the appropriate SAFE model, elastodynamic fields in the straight and in the curved parts are expressed as modal series of guided modes traveling in opposite directions as shown by Fig. 2 and in the following equation,

$$\begin{pmatrix}\mathbf{X}\\ \mathbf{Y}\end{pmatrix}=\sum_n A_n\begin{pmatrix}\mathbf{X}_n^+\\ \mathbf{Y}_n^+\end{pmatrix}e^{-j\beta_n^+ z}+\sum_n B_n\begin{pmatrix}\mathbf{X}_n^-\\ \mathbf{Y}_n^-\end{pmatrix}e^{-j\beta_n^- z} \tag{7}$$

where $\mathbf{X} = (\sigma_{xy}, \sigma_{yz}, u_z)^T$ and $\mathbf{Y} = (u_x, u_y, -\sigma_{zz})^T$ are the hybrid vectors introduced by [7] and coupling displacement and stress components of the elastodynamic field, $\mathbf{X}_n$ and $\mathbf{Y}_n$ are the hybrid vectors of each eigenmode obtained using SAFE with $\mathbf{X}_n^- = -\mathbf{X}_n^+$, $\mathbf{Y}_n^- = \mathbf{Y}_n^+$, $\beta_n^- = -\beta_n^+$, A_n and B_n represent the amplitudes of right going and the left going modes in each guide.

At the junction, the two elastodynamic fields associated to the different parts of the pipework and expressed by Eq. (7) must satisfy the continuity condition. In the present case, the elastodynamic fields are those associated to the straight (superscript 1) and the curved (superscript 2) parts of the pipework. Then, the continuity is written as:

$$\begin{gathered}\begin{pmatrix}\mathbf{X}^1\\ \mathbf{Y}^1\end{pmatrix}=\begin{pmatrix}\mathbf{X}^2\\ \mathbf{Y}^2\end{pmatrix}\quad at\ z=s=z_0\\ \sum_{n_1}A_{n_1}^1\begin{pmatrix}\mathbf{X}_{n_1}^1\\ \mathbf{Y}_{n_1}^1\end{pmatrix}e^{-j\beta_{n_1}^1 z}+\sum_{n_1}B_{n_1}^1\begin{pmatrix}-\mathbf{X}_{n_1}^1\\ \mathbf{Y}_{n_1}^1\end{pmatrix}e^{+j\beta_{n_1}^1 z}=\sum_{n_2}A_{n_2}^2\begin{pmatrix}\mathbf{X}_{n_2}^2\\ \mathbf{Y}_{n_2}^2\end{pmatrix}e^{-j\beta_{n2}^2 s}+\sum_{n_2}B_n^2\begin{pmatrix}-\mathbf{X}_{n_2}^2\\ \mathbf{Y}_{n_2}^2\end{pmatrix}e^{+j\beta_{n2}^2 s}\end{gathered} \tag{8}$$

Following Feng *et al.* [8] and using the biorthogonality relation (Fraser's relation), the mode matching method expresses the relation between the amplitudes of the outgoing and the incoming modes by projecting Eq. (8) on the modal basis calculated by SAFE. Finally, one gets the following expression for the scattering matrix **S** at the junction:

$$\begin{gathered}\begin{pmatrix}\mathbf{B}_1\\ \mathbf{A}_2\end{pmatrix}=\mathbf{S}\begin{pmatrix}\mathbf{A}_1\\ \mathbf{B}_2\end{pmatrix}\\ \mathbf{S}=\begin{pmatrix}\mathbf{R}_{1\to2} & \mathbf{T}_{2\to1}\\ \mathbf{T}_{1\to2} & \mathbf{R}_{2\to1}\end{pmatrix}=\begin{pmatrix}2\mathbf{J}_1 & \mathbf{J}_{2,1}^T-\mathbf{J}_{1,2}\\ \mathbf{J}_{1,2}^T-\mathbf{J}_{2,1} & 2\mathbf{J}_2\end{pmatrix}^{-1}\begin{pmatrix}[\mathbf{0}]_{N_1\times N_2} & \mathbf{J}_{1,2}+\mathbf{J}_{2,1}^T\\ \mathbf{J}_{2,1}+\mathbf{J}_{1,2}^T & [\mathbf{0}]_{N_2\times N_1}\end{pmatrix},\end{gathered} \tag{9}$$

where $\mathbf{R}_{1\to2}$, $\mathbf{T}_{1\to2}$ ($\mathbf{R}_{2\to1}$, $\mathbf{T}_{2\to1}$) are the reflection and transmission matrices for incident modes in the straight part (respectively, in the curved part). N_i (i = 1, 2) denote the numbers of modes considered for applying the mode matching in each part of the guide. All the propagative modes in both parts are taken into account together with a number of evanescent modes for reducing numerical errors. $\mathbf{J_i}$ and $\mathbf{J_{i,j}}$ are the projection matrices of the elastodynamic fields calculated with the bi-orthogonality relation. Their elementary components for the n-th mode in guide 1 onto the m-th mode in guide 1 [Eq. (10a)] or in guide 2 [Eq. (10b)] are defined as:

$$J_i(n,m) = (\mathbf{X}_n^i | \mathbf{Y}_m^i) = \int_\Sigma \mathbf{X}_n^i(x,y)\mathbf{Y}_m^i(x,y)\sqrt{\det(\mathbf{g}_i)}\, dxdy = \delta_{n,m}\, \mathbf{X}_n^i \mathbf{M}_s^i \mathbf{Y}_m^i, \quad (10a)$$

$$J_{i,j}(n,m) = (\mathbf{X}_n^i | \mathbf{Y}_n^j) = \int_\Sigma \mathbf{X}_n^i(x,y)\mathbf{Y}_n^j(x,y)\sqrt{\det(\mathbf{g}_j)}\, dxdy = \mathbf{X}_n^i \mathbf{M}_s^j \mathbf{Y}_n^j. \quad (10b)$$

$\mathbf{M}_s^i$ (i =1, 2) are the finite element projection matrices associated with the straight and the curved guides. These matrices are obtained from the global mass matrices calculated by the SAFE models and differ in each guide.

2.3. Expression of the global scattering matrix

As shown previously, the mode matching method computes the scattering matrix of the junction by predicting the local phenomena linking two different elastodynamic fields computed by SAFE. However, pipeworks generally comprise a succession of junctions (similar or dissimilar). At each junction, various GW scattering phenomena occur which can be translated mathematically in the form of a specific modal scattering matrix. It is of course interesting to develop a way of predicting the overall GW propagation through a series of scatterers, similarly translated mathematically as a global matrix. Such an overall formulation of general applicability has been recently developed by two of the present authors [9]. In this formulation, a modal scattering matrix is obtained by cascading local scattering matrices (junctions, defects, etc.) and modal matrices accounting for the GW propagation back and forth in between.

In this section, this theory initially derived for straight guiding portions is adapted to deal with locally curved parts, our aim being to deal with the GW scattering by an elbow joined to two straight tubes.

The global scattering matrix must deal with the multiple reflections between the two junctions and associated transmissions by linking the amplitudes of incoming modes to those of outgoing modes of the two straight parts of the pipework as illustrated in Fig 3:

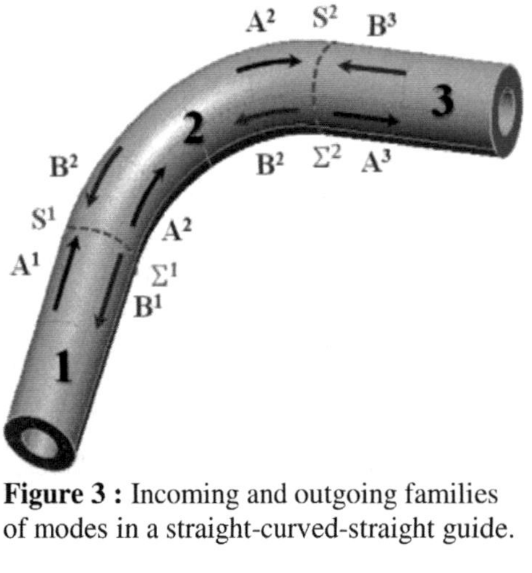

Figure 3 : Incoming and outgoing families of modes in a straight-curved-straight guide.

According to [9], the global scattering matrix is a combination of the local scattering matrices S_1 and S_2 at the two junctions and the propagation matrix of the modes in the elbow. This latter is easily adapted here to the case of a curved guide by using the curvilinear length $s_c = R_c\,\theta_c$ of the centreline of the elbow; it writes:

$$\mathbf{P}^{\text{elbow}} = \mathbf{diag}(e^{i\beta_n s_c})_{1\le n\le N_2} \tag{11}$$

and the global scattering matrix $\mathbf{S}_G$ is given by:

$$\begin{pmatrix}\mathbf{B}^1\\ \mathbf{A}^3\end{pmatrix} = \mathbf{S}_G\begin{pmatrix}\mathbf{A}^1\\ \mathbf{B}^3\end{pmatrix} \text{with}\,\mathbf{S}_G = \begin{pmatrix}\mathbf{R}_{G_1} & \mathbf{T}_{G_{31}}\\ \mathbf{T}_{G_{13}} & \mathbf{R}_{G_3}\end{pmatrix} \tag{12}$$

From this expression of $\mathbf{S}_G$, it is easy to extract the modal decomposition of incident modes in either one of the straight parts as reflected or transmitted modal components in the same part or in the other part. Reflection and transmission sub-matrices in Eq. (12) are given by [9]:

$$\begin{aligned}
\mathbf{R}_{G_1} &= \mathbf{R}^{(S_1)}_{1\to2} - \mathbf{T}^{(S_1)}_{2\to1}(\mathbf{I}-\boldsymbol{\Lambda}^{(1\to3)})^{-1}(\mathbf{R}^{(S_1)}_{2\to1})^{-1}\mathbf{T}^{(S_1)}_{1\to2},\\
\mathbf{T}_{G_{13}} &= -\mathbf{T}^{(S_2)}_{3\to2}[(\mathbf{I}-\boldsymbol{\Lambda}^{(3\to1)})^{-1}\boldsymbol{\Lambda}^{(3\to1)}\mathbf{P}^{\text{elbow}}\mathbf{T}^{(S_1)}_{1\to2}],\\
\mathbf{R}_{G_3} &= \mathbf{R}^{(S_1)}_{3\to2} - \mathbf{T}^{(S_2)}_{2\to3}(\mathbf{I}-\boldsymbol{\Lambda}^{(1\to3)})^{-1}(\mathbf{R}^{(S_1)}_{2\to3})^{-1}\mathbf{T}^{(S_2)}_{3\to2},\\
\mathbf{T}_{G_{31}} &= -\mathbf{T}^{(S_2)}_{3\to2}[(\mathbf{I}-\boldsymbol{\Lambda}^{(3\to1)})^{-1}\boldsymbol{\Lambda}^{(3\to1)}\mathbf{P}^{\text{elbow}}\mathbf{T}^{(S_2)}_{3\to2}],\\
\text{with}\ \boldsymbol{\Lambda}^{(1\to3)} &= (\mathbf{R}^{(S_1)}_{2\to1})^{-1}(\mathbf{P}^{\text{elbow}})^{-1}(\mathbf{R}^{(S_2)}_{2\to3})^{-1}(\mathbf{P}^{\text{elbow}})^{-1},\\
\text{and}\ \ \boldsymbol{\Lambda}^{(3\to1)} &= (\mathbf{R}^{(S_2)}_{2\to3})^{-1}(\mathbf{P}^{\text{elbow}})^{-1}(\mathbf{R}^{(S_1)}_{3\to2})^{-1}(\mathbf{P}^{\text{elbow}})^{-1}.
\end{aligned} \tag{13}$$

The notations $\mathbf{R}^{(S_i)}_{j\to k}$ and $\mathbf{T}^{(S_i)}_{j\to k}$ have an obvious meaning of reflection and transmission matrices related to the two junctions. The same approach is straightforwardly generalized for computing the global scattering matrix of a pipework comprising several elbows with different curvatures by cascading the appropriate scattering and propagation matrices.

3. An illustrative example of GW scattering by an elbow

Consider an incident T(0,1) mode (commonly used in NDT operations in pipeworks) in a straight-curved-straight guide with an elbow of 120° as shown in Figure 4.

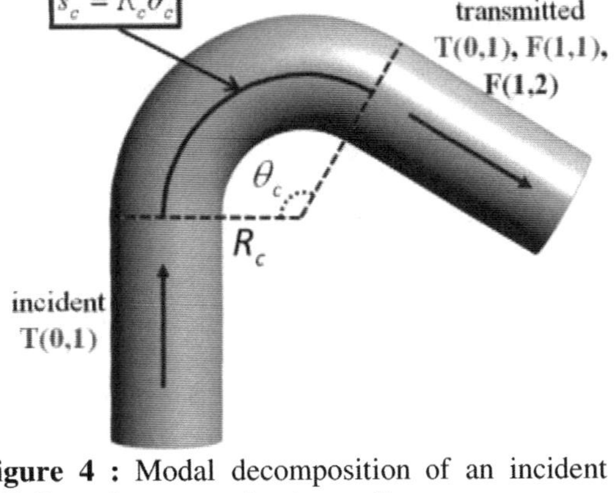

Figure 4 : Modal decomposition of an incident T(0,1) mode propagating in an elbow.

The energy ratios of an incident T(0,1) among transmitted modes at a single straight/curved junction and in an entire elbow as a function of the frequency are shown on Figure 5 (a) and (b):

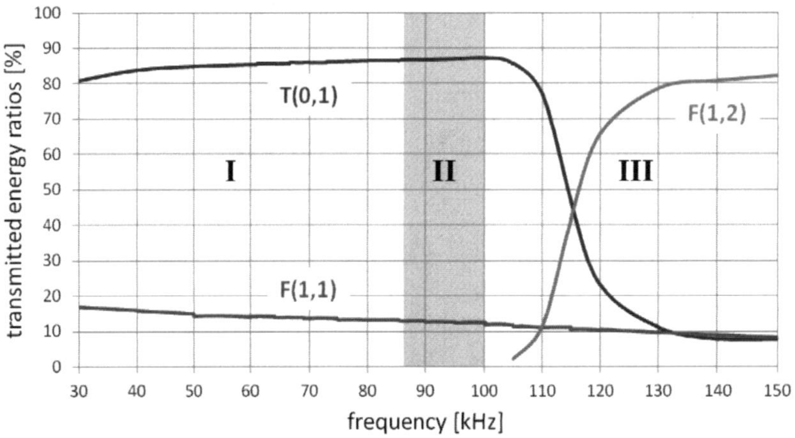

Figure 5(a) : Transmitted energy for an incident T(0,1) mode at a single junction between a straight and a curved guide

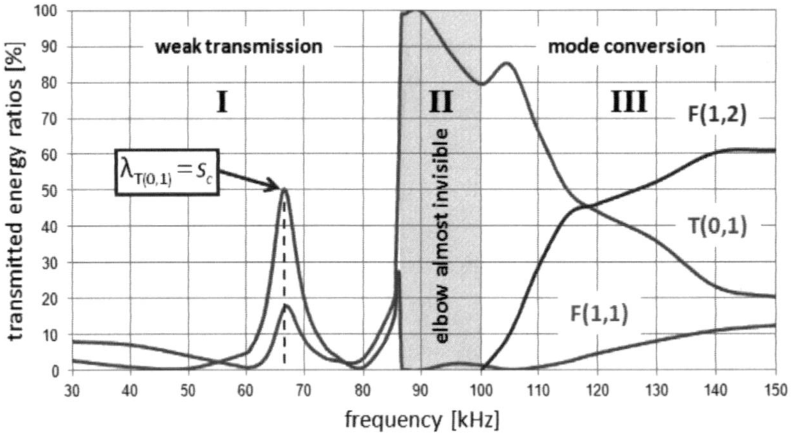

Figure 5(b) : Transmitted energy for an incident T(0,1) mode vs. frequency.

Three frequency bands can be distinguished. The first one ([0-87] kHz) is characterized by a weak transmission and weak mode conversions considering the entire elbow despite the fact that the incident mode is highly transmitted at the first junction of the guide, except around the frequency for which the value of the T(0,1) wavelength equals the length of elbow centerline. In the second band ([87-100] kHz), the transmission still very high at the first junction and the entire elbow becomes almost transparent (high transmission ratio) with a very weak mode conversion; such a frequency range could be exploited for inspecting long pipeworks comprising several similar elbows. Finally, in the third band ([100-150] kHz), increasing mode conversion occurs at the single junction as well as in the entire elbow, notably into a high amplitude transmitted flexural mode F(1,2), though energy transmitted as a T(0,1) mode is still important at low frequencies of this band but decreases as the frequency increases.

This result clearly shows that the chosen working frequency range is an important criterion which directly influences the guided waves behaviour when GW propagation in pipeworks comprising elbows is considered. These curves can be very useful for industrial inspections; in practice, it can help to identify the appropriate frequency range to get the most efficient transmission or reflection phenomena and to anticipate modes conversions at each junction of the considered pipeworks and multiple reflections between junctions.

4. Experimental investigation of guided wave propagation in an industrial elbow

4.1. Experimental set-up

An experimental study was performed in collaboration with the Cetim (Technical Center for the mechanical industry) in order to investigate the effect of curvature on the guided waves propagation in industrial pipeworks comprising elbows. The experimental model consists in a 6 inches and a 7.1mm thickness steel pipe with an elbow of 30° and 437mm bend radius joined to two straight parts as shown in Figure 6:

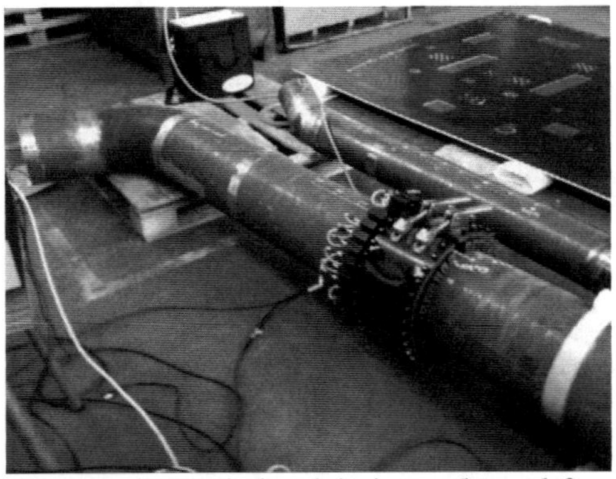

Figure 6: The industrial pipework used for the experimental study.

Figure 7 illustrate the different features of our experimental model and their position:

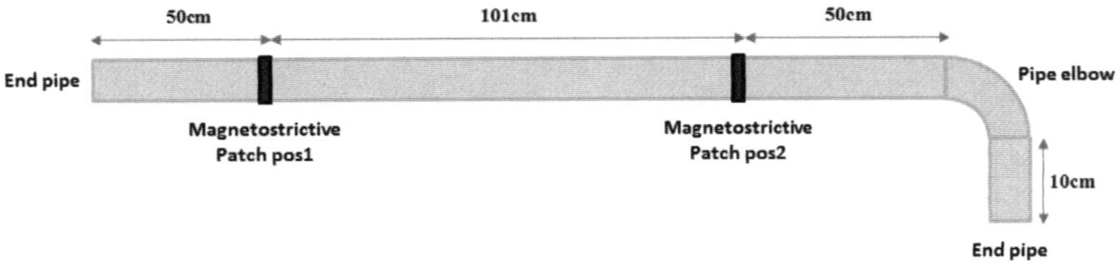

Figure 7: Schematic view of the experimental model.

The data acquisition device consists of the MsS System [10] produced by Guided Wave Analysis LLC (GWA) and a point-like piezoelectric (Teletest) shear wave receiver produced by Plant integrity LTD (Pi) [12]. The MsS system is a magnetostrictive device used to generate and receive low-frequency (20 kHz to 250 kHz) guided waves (T(0,1) and L(0,2)) for NDT operations.

In our case, the system is exclusively used as a guided waves generator in a pulse/echo configuration (The emitter and the receiver are combined in a unique MsS patch) to generate the first torsional mode T(0,1) which is the only non-dispersive propagative mode (see Figure 8(a)).

The point-like Teletest sensor (see figure 8(b)) is an instrumented shear wave sensor coupled to the pipe with a shear gel and used as receiver. The experimental set-up is also composed of a digital oscilloscope to record filtered and amplified signals received.

Figure 8(a): The magnetostrictive patch (MsS system) used to generate the guided waves.

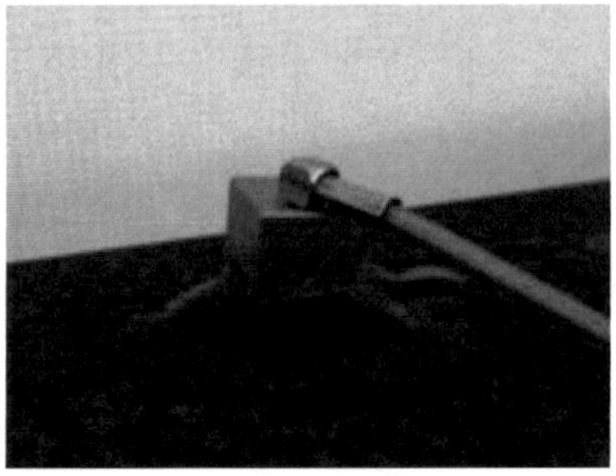

Figure 8(b): Point-like piezoelectric element (Teletest) used as a receiver coupled with a shear gel.

4.2. Experimental acquisitions in different configurations

The MsS System is used in a pulse/echo configuration at the patch located at position 1 (see Figure 7) to generate the first torsional mode T(0,1). In this configuration, received signals measured are typical of that shown on Figure 9, for a 2 cycle Hanning-windowed excitation centred at 64 kHz:

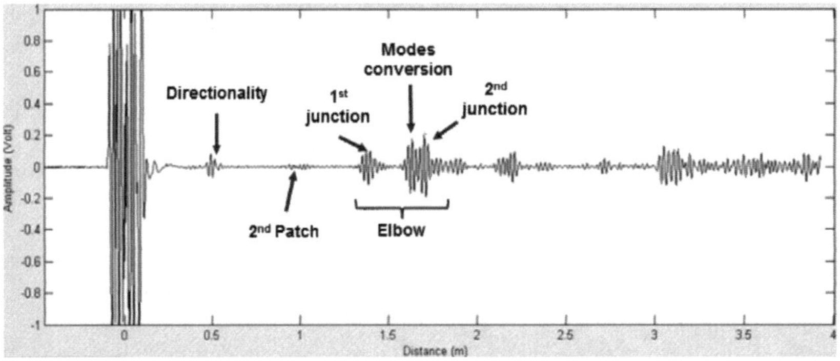

Figure 9: A-scan obtained with the MsS system in a pulse/echo configuration

The A-scan shows the echoes reflected from various features of the pipework. The first one comes from the first junction between the straight and the curved pipe without mode conversion (the T(0,1) incident pulse is reflected as a T(0,1) mode). The second one is due to mode conversions of the incident torsional mode as flexural modes in the elbow, which re-convert into a torsional mode; the third echo comes from the reflection at the second junction of the incident T(0,1) in itself without mode conversion.

The second acquisition is based on the use of the instrumented point-like piezoelectric element used as a receiver. The incident mode is again generated by the magnetostrictive patch located at position 1. By contrast to the reception using the MsS system, the point-like sensor with a sensitivity in the plane perpendicular to the guide axis is thus sensitive to the orthoradial displacement component of flexural modes generated by mode conversion in the elbow.

With this setup, flexural and torsional modes are detected; modes can be identified and classified by estimating their wavespeed in the pipe. For example, by looking at the echoes arising from the elbowed part, an echo generated by mode conversion at the first junction arriving after the echo from the first junction is visible. It cannot be detected by the MsS system in pulse-echo configuration because summation of associated displacement over the sensor area, that is to say, over the entire pipe circumference, cancels the contributions of flexural modes.

In addition, by performing successive acquisitions using the point-like piezoelectric sensor with a measure step of 0.5 cm at 128 kHz and 1 cm at 32 kHz and 64 kHz, a B-scan-like image can be formed by superimposing all the (A-scans) signals recorded at the various positions. This allows us identifying the wave packets and estimating wave velocities of the incident and reflected modes propagating in the pipework. Figure 10 shows an example of such a B-scan built by superimposing 30 A-scans measured by the point-like sensor on the 6 inches pipework at a central frequency of 64 kHz.

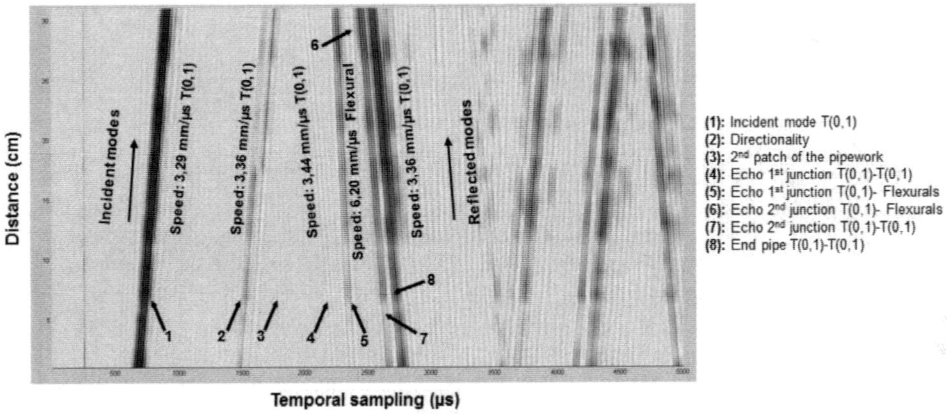

Figure 10: B-scan plotted by superimposing the successive Ascans acquisitions obtained using the point-like piezoelectric sensor

Wave packets in Figure 10 represent the incident and the reflected modes in the pipework. Modes with a positive slope correspond to incident modes; those with a negative slope are the reflected ones. Wavespeeds are directly measurable from the figure and the nature of modes (flexural or torsional) can be identified. For example, the 6th wave packet in Fig. 10 propagates at 6.2 mm/µs which cannot correspond to the velocity of the torsional mode emitted (3.24 mm/µs); it is a flexural mode created by mode conversion at the junction of the elbow. The polarization of the point-like piezoelectric sensor used is makes it otherwise insensitive to extensional modes.

4.3. Study of the modes conversion in the elbow

The same dataset can also be post-processed to better identify the various modes to which the sensor is sensitive generated at the two junctions of the elbow. This post-processing consists in computing a two-dimensional fast Fourier transform (FFT2D), allowing us to separate propagative modes by identifying their respective wavenumbers over the frequency spectrum of the excitation signal. Here, three excitation signals were used, made of two cycles of a sine of centre frequencies 32 kHz, 64 kHz and 128 kHz filtered by a Hanning-window. Combining the three broad bandwidths allows us to follow wavenumber variations over a frequency range from 5 kHz to 175 kHz (Figure 11):

On Fig. 11, the curve with a positive slope corresponds to the incident mode generated by the MsS system; this confirms that the first torsional mode T(0,1) is the only mode excited using this system. Curves with negative slopes correspond to reflected T(0,1) mode together with many modes created by mode conversion arising from the scattering of the incident mode at the two junctions of the elbow. This experimental result can be easily compared to the dispersion curves computed for a straight tube. An excellent fit was observed by superposing such dispersion curves onto results shown on Fig. 11 but it is rather hard to display graphically on the same figure for the present paper.

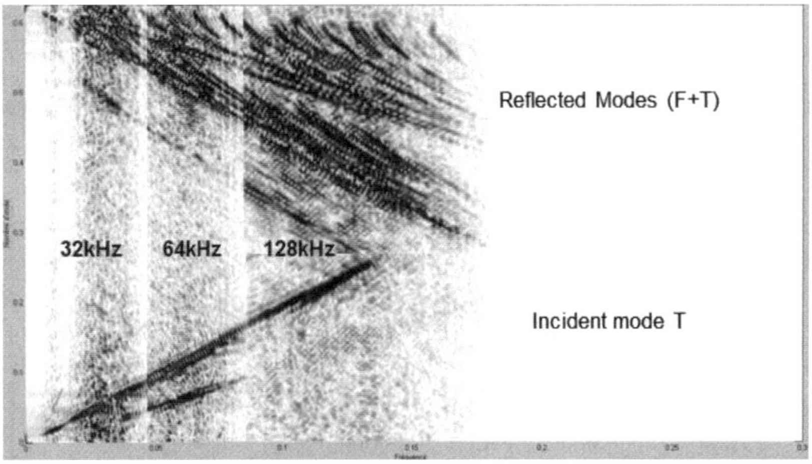

Figure 11: FFT2D of the incident mode T(0,1) and the reflected modes generated by the modes conversions in the elbow.

4.4. Comparison between numerical simulation and experimental signals

In pipe inspections by guided waves, A-scans are used to get indications about the integrity of the pipework. By interpreting this time-domain signal, information can be extracted such as the presence or not of defects and their locations, mode conversions at junctions, reflection and/or transmission coefficients and times of flight…

Here, the comparison of measured and computed A-scans is made to demonstrate the model ability to predict the complex behaviour of guided waves as they propagate and are scattered into an industrial pipework comprising an elbow. This comparison is given by figure 12:

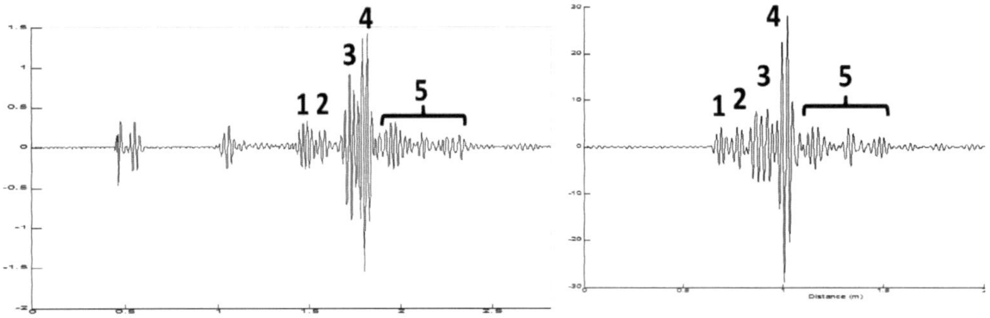

Figure 12: comparison between experimental (blue) and simulated (black) A-scans of an inspection using T(0,1) of a pipework of 6 inches pipes comprising an elbow of 30° at 64kHz

The different echoes correspond to:

(1) the reflection at the entry of the elbow **T(0,1)-T(0,1)**

(2) a series of mode conversions resulting from the initial mode conversion of the incident T(0,1) mode into flexural modes re-converted as torsional one, denoted by **T(0,1)-F-T(0,1)**

(3) reflection from the exit of the elbow of the torsional mode in the elbow similar to the T(0,1) incident mode denoted by **T(0,1)-T(0,1)**

(4) reflection at the end pipe **T(0,1)-T(0,1)**

(5) multiple scattering phenomena between the two junctions.

The comparison of the two signals shows a very good agreement in terms of number, times of flight and amplitudes of the echoes despite uncertainties with regard to elbow characteristics (irregular curvature, wavy surface resulting from the bending process…). The multiple scattering phenomena (echoes 5) occurring after the echo of the end pipe are due to reflection and transmission between the latter and the second junction of the elbow. Further experimental validations are in progress to confirm the capabilities of the proposed model to predict these complex scattering phenomena.

5. Conclusion

A numerical model based on a modal approach and using a SAFE method coupled to a mode matching technique allows predicting the propagation and the scattering of guided waves into pipeworks comprising elbows. A single overall scattering matrix can be computed by combining propagation matrices to local scattering matrices. Interestingly, all the computations only imply meshing the guide cross-sections and solving a system of equations (FE) is only required to compute the modal basis in the various guiding parts involved (straight and curved). Thus, the overall computation allows the prediction of guided waves scattering phenomena at a low cost in terms of time and computer resources. An example of numerical parametric study of the transmission through an elbow upon frequency has been made and shows some complex scattering behaviour which highlights the importance of simulation tools for choosing the appropriate frequency for optimal inspection. Finally, an experimental study has been performed to confirm these complex diffraction phenomena produced by elbows in industrial pipeworks. Further numerical and experimental validations will be performed before integrating the developed model into the simulation software platform CIVA [3].

References

[1] T. Hayashi, K. Kawashima, Z. Sun and J.L. Rose, *J. Press.Vessel Tech.* **127**, 322-327 (2005).
[2] A. Demma, P. Cawley, M. Lowe and B. Pavlakovic, *J. Press.Vessel Tech.* **127,** 328-335 (2005).
[3] General information about the CIVA software platform can be found at www-civa.cea.fr
[4] M. El Bakkali, A. Lhémery, V. Baronian and F. Berthelot, *AIP Conf. Proc.*, **1581**, 332, (2014).
[5] D. Chapelle and K.-J. Bathe, *The finite element analysis of shells – Fundamentals*, Springer, Berlin, 2003.
[6] F. Treyssède and L. Laguerre, *J. Sound Vib.* **329**, 1702-1716, (2010).
[7] V. Pagneux and A. Maurel, *Proc. R. Soc.-A* **426**, 1315-1339 (2006).
[8] F. Feng, J. Shen and S. Lin, *Ultrasonics* **52**, 760-766 (2012).
[9] V. Baronian, A. Lhémery and K. Jezzine, "Hybrid SAFE/FE simulation of inspections of elastic waveguides containing several local discontinuities or defects," in *Review of Progress in QNDE*, ed. by D.O. Thompson and D. E. Chimenti, AIP Conference Proceedings 1211, American Institute of Physics, Melville, NY, 2010, pp. 183-190.
[10] Guided Wave Analysis LLC website : http://www.gwanalysis.com/index.html
[11] Plant Integrity Ltd website : http://www.plantintegrity.com/index.jsp

Propagation number in periodic structures considering losses

P Méresse[1], Y Renou[1], J Vasseur[3], C Audoly[1] and A C Hladky[2]

[1]DCNS Research, BP 403 Le Mourillon 83055 Toulon Cedex, FRANCE
[2]IEMN, ISEN department, UMR 8520 CNRS, Lille, FRANCE
[3]IEMN, UMR 8520 CNRS, Villeneuve dAscq, FRANCE

E-mail: pierre.meresse@gmail.com

Abstract. The development of absorbing or reflecting panels is of interest for underwater acoustic stealth. Most of the time, losses are not considered in the structures studied, specifically in the propagation number determination, despite the fact that most materials used, such as polymer matrices, have non negligible viscous losses. So, for a better understanding of the acoustic properties of these structures and to allow their optimization, simulations should consider the material properties, including losses. In order to obtain more realistic results from simulation, two numerical tools based on the finite element method (FEM) are proposed, with the help of the *ATILA* software. One is based on a differential method, the other on the transfer matrix. The two methods are first validated in the lossless case, then when losses are taken into account. Both methods give results in good agreement and give the propagation number where losses are taken into account.

1. Introduction

Since the Second World War, acoustic control is a priority in naval stealth. Ships need absorbing or reflecting panels. Periodic materials are a potential solution for this problem. These structures behave as mechanical filters which do not allow wave propagation in certain frequency ranges. Periodic structure characterization is done by resolving the dispersion relation where the angular frequency ω is a function of the wave number k that is $\omega = f(k)$. This computation usually does not consider damping terms in materials. However, acoustic panels are made of materials with non negligible viscous losses in order to increase absorption.

There are several ways to compute the dispersion curve considering periodic structures. Gaunaurd [1] presented a one-dimensional analysis of cylindrical cavities, Audoly [2] compared his work with the Waterman-Truell theory about random distribution of inclusions. Wei [3] worked on the influence of viscosity and he determined band gaps with the plane wave expansion method. Here the finite element method (FEM) provided by the *ATILA* software [4, 5] is used. Thanks to this tool, the modal analysis of a meshed structure gives real solutions for the dispersion relation [6, 7]. This computation considers materials without losses. However, the software is able to consider viscosity during the harmonic analysis, transmission and reflection coefficients can be calculated for a slab with a finite thickness. In this paper, two different methods are presented : Bianco & Parodi method, which is a differential method, and transfer matrix method. They have been adapted to retrieve the dispersion relation considering losses in the materials. Both methods are able to take into account damping due to the materials used.

An example of a periodic bi-layer structure is studied. In a first step, no damping is considered to validate the methods. Next, a small part of viscous losses is added into the material in order to observe the damping influence on the dispersion curve.

2. Transfer matrix method

This method proposes to compute all complex solutions for the wave number at a given frequency. It has first been developed by D. J. Mead [8, 9] and applied by M. L. Accorsi [10] in periodic structures. Lately, it was used by M. Bavencoffe [11].

For a given frequency ω_0, finite element equation defines a relation between the nodal displacement $\tilde{U}$ and the nodal applied force $\tilde{F}$ with the help of the stiffness matrix K and the mass matrix M. This relation connecting $\tilde{U}$ and $\tilde{F}$ can be summarised as written in equation 1,

$$\left([K] - \omega_0^2[M]\right)\tilde{U} = [D]\tilde{U} = \tilde{F}. \tag{1}$$

Moreover, positions can be discriminated between the left side nodes (l), right side nodes (r) and inner nodes (i) of the elementary cell, as shown on figure 1 which is an example of a meshed multi-layer cell. Equation 2 considers this distinction, written

$$\begin{bmatrix} D^{ll} & D^{li} & D^{lr} \\ D^{il} & D^{ii} & D^{ir} \\ D^{rl} & D^{ri} & D^{rr} \end{bmatrix} \begin{pmatrix} \tilde{U}^l \\ \tilde{U}^i \\ \tilde{U}^r \end{pmatrix} = \begin{pmatrix} \tilde{F}^l \\ \tilde{F}^i \\ \tilde{F}^r \end{pmatrix}. \tag{2}$$

The free waves condition implies that no force is applied on inner nodes, i.e. $\tilde{F}_i = 0$. This condition allows to retrieve the expression for inside nodes displacement as follows :

$$\tilde{U}_i = -[D^{ii}]^{-1}(D^{il}\tilde{U}^l + D^{ir}\tilde{U}^r). \tag{3}$$

If $\tilde{U}_i$ is replaced in equation 2, it can be written as

$$\begin{bmatrix} D^{ll} - D^{li}[D^{ii}]^{-1}D^{il} & D^{lr} - D^{li}[D^{ii}]^{-1}D^{ir} \\ D^{rl} - D^{ri}[D^{ii}]^{-1}D^{il} & D^{rr} - D^{ri}[D^{ii}]^{-1}D^{ir} \end{bmatrix} \begin{pmatrix} \tilde{U}^l \\ \tilde{U}^r \end{pmatrix} = \begin{bmatrix} A & B \\ C & D \end{bmatrix} \begin{pmatrix} \tilde{U}^l \\ \tilde{U}^r \end{pmatrix} = \begin{pmatrix} \tilde{F}^l \\ \tilde{F}^r \end{pmatrix}, \tag{4}$$

where the left matrix is summarized by A, B, C and D terms for a tractable equation. As a periodic structure, the $\jmath^{th}$ cell is considered. Also, $\jmath^{th}$ cell's right side is $(\jmath+1)^{th}$ cell's left side. In other terms, if $U^l = U^\jmath$, $U^r = U^{\jmath+1}$. Using this notation, equation 4 is equivalent to

$$\begin{bmatrix} A & B \\ C & D \end{bmatrix} \begin{pmatrix} \tilde{U}^\jmath \\ \tilde{U}^{\jmath+1} \end{pmatrix} = \begin{pmatrix} \tilde{F}^\jmath \\ -\tilde{F}^{\jmath+1} \end{pmatrix}. \tag{5}$$

Now, a relation can be defined between the quantities of interest at the right hand side of the cell and those at the other side,

$$\begin{pmatrix} \tilde{U}^{\jmath+1} \\ \tilde{F}^{\jmath+1} \end{pmatrix} = \begin{bmatrix} -B^{-1}A & B^{-1} \\ DB^{-1}A - C & -DB^{-1} \end{bmatrix} \begin{pmatrix} \tilde{U}^\jmath \\ \tilde{F}^\jmath \end{pmatrix} = [T] \begin{pmatrix} \tilde{U}^\jmath \\ \tilde{F}^\jmath \end{pmatrix}. \tag{6}$$

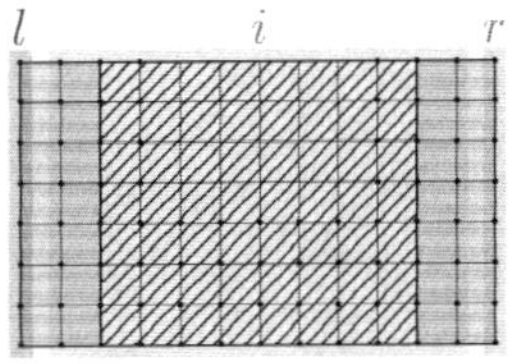

Figure 1. Mesh for an elementary cell of symetric multi-layer structure - Plain : layer 1 ; stripped lines : layer 2

A similar relation exists in periodic medium. It is the Bloch-Floquet relation, using the wave number k and periodic distance a,

$$\begin{pmatrix} \tilde{U}^{J+1} \\ \tilde{F}^{J+1} \end{pmatrix} = e^{jka} \begin{pmatrix} \tilde{U}^{J} \\ \tilde{F}^{J} \end{pmatrix}. \tag{7}$$

Both equations 6 and 7 are combined to get a dispersion relation :

$$[T - e^{jka} I] \begin{pmatrix} \tilde{U}^{J} \\ \tilde{F}^{J} \end{pmatrix} = 0. \tag{8}$$

3. Bianco & Parodi method

This method was firstly developed to find out electromagnetic waves in micro-strip. This task is difficult due to interface effects and evanescent waves. B. Bianco and M. Parodi [12] came with the idea of comparing measurements from two lines with different lengths. Both lines are long enough to allow only the first order mode in the middle part. The aim is to retrieve the wave number in the material core. Lately, this method was applied by C. Croenne [13] in metamaterials.

On figure 2, two slabs made of the same medium but with some extra-material Δl added in the second case, between points 5 and 6, are represented.

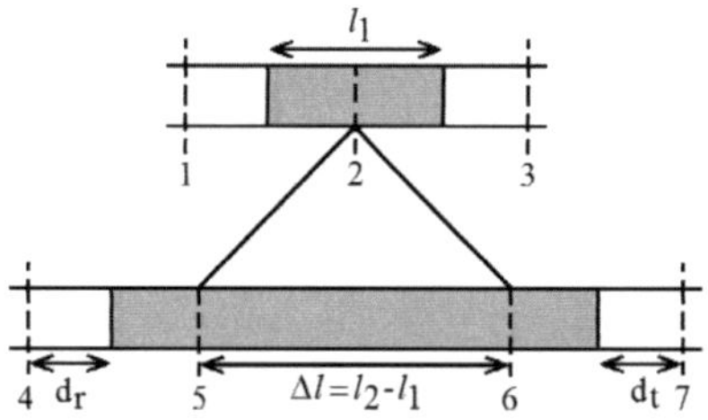

Figure 2. Drawing of two different slabs (grey) with a variation of thickness Δl surrounded by fluid (white)

The matrix representation is considered for the case presented on figure 2. M_{ij} is the propagation matrix between points i and j. Some relations appear :

$$M_{13} = M_{12} M_{23} \quad ; \quad M_{47} = M_{45} M_{56} M_{67} = M_{12} M_{56} M_{23}. \tag{9}$$

Now, the product of the second slab matrix with the inverse matrix of the first slab,

$$M_{47} M_{13}^{-1} = (M_{12} M_{56} M_{23}) M_{23}^{-1} M_{12}^{-1} = M_{12} M_{56} M_{12}^{-1} \tag{10}$$

is computed. One can readily conclude that $M_{47} M_{13}^{-1}$ and M_{56} are similar matrices. So traces of these matrices are identical, i.e.

$$tr(M_{47} M_{13}^{-1}) = tr(M_{56}). \tag{11}$$

M_{56} can be expressed using the expression from transmission lines theory for Δl length,

$$M_{56} = \begin{bmatrix} \cos(k\Delta l) & jZ_c \sin(k\Delta l) \\ jZ_c^{-1} \sin(k\Delta l) & \cos(k\Delta l) \end{bmatrix}. \tag{12}$$

The wave number k appears. Then the first part of equation 11 is expressed using D. A. Frickey's work on S parameters [14]. Frickey shows that the propagation matrix is expressed

$$M = \frac{1}{2S_{21}} \begin{bmatrix} (1+S_{11})(1-S_{22})+S_{12}S_{21} & (1+S_{11})(1+S_{22})-S_{12}S_{21} \\ (1-S_{11})(1-S_{22})+S_{12}S_{21} & (1-S_{11})(1+S_{22})+S_{12}S_{21} \end{bmatrix}, \tag{13}$$

which can be simplified. If slabs are reciprocal, the reflection coefficient is noted $r = S_{11} = S_{22}$ and transmission $t = S_{12} = S_{21}$. Coefficients are independent of the propagation way, *e.g.* with a symetrical disposition. The propagation matrix can be expressed with these coefficients as follows :

$$M_{13} = \frac{1}{2t_1} \begin{bmatrix} 1-r_1^2+t_1^2 & (1+r_1)^2-t_1^2 \\ (1-r_1)^2-t_1^2 & 1-r_1^2+t_1^2 \end{bmatrix}; \tag{14a}$$

$$M_{47} = \frac{1}{2t_2} \begin{bmatrix} 1-r_2^2+t_2^2 & (1+r_2)^2-t_2^2 \\ (1-r_2)^2-t_2^2 & 1-r_2^2+t_2^2 \end{bmatrix}, \tag{14b}$$

where subscripts indicate the thin slab (1) or the thick slab (2). Each term r_i and t_i is associated to a delay term $e^{jk_0 d}$ due to propagation in the fluid distance d_r or d_t (figure 2) with k_0 as the wave number in the surrounding fluid. All delay terms are summarized in a unique exponential term in equation 15, which is the pseudo dispersion relation from equation 11,

$$2\cos(k\Delta l) = \frac{t_1^2 + t_2^2 - (r_1 - r_2)^2 e^{2jk_0(d_t - d_r)}}{t_1 t_2}. \tag{15}$$

4. Results

4.1. Without losses

In order to validate the tools developed, a periodic multi-layer material is numerically studied. The elementary cell of the structure is made of 60 % steel and 40 % epoxy with a periodic distance a equal to 10mm. The disposition is symetrical, as shown in figure 1, to satisfy equations 14 conditions. This case is simple enough to allow an analytical resolution in order to compare results. Analytical solutions for longitudinal waves and transverse waves are developed in Langlet's work [6]. Dispersion relation is

$$\cos(ak) = \cos(\frac{\omega d_1}{c_1})\cos(\frac{\omega d_2}{c_2}) - \frac{(\rho_1 c_1)^2 + (\rho_2 c_2)^2}{2\rho_1\rho_2 c_1 c_2}\sin(\frac{\omega d_1}{c_1})\sin(\frac{\omega d_2}{c_2}) \tag{16}$$

where different materials are subscripted 1 or 2, with ρ the volumic density, c the wave celerity (longitudinal or tranverse) and d the material thickness where $d_1 + d_2 = a$. In a first step, the quality of results need to be ensured. So computations are considered without losses. Results are presented on figure 3. To allow easy reading, analytical solutions are represented by lines on both figures. Longitudinal waves colored in red are distinguished from transverse waves colored in blue. Results are represented by dots, computed by the transfer matrix method in figure 3a and computed by the Bianco & Parodi method in figure 3b. Both methods give results in good agreement with analytical solution. Here the wave number gets an imaginary part, contrary to the modal analysis by FEM that only gives the real part [7]. It helps the dispersion curve reading by following the modes branches. One can notice that there is less information with Bianco & Parodi results. The reflection and transmission coefficients used are computed for a normal incident wave in a fluid environment. In these conditions, only information about longitudinal modes can be retrieved. This particularity allows to identify clearly which curve corresponds to a longitudinal or a transverse mode.

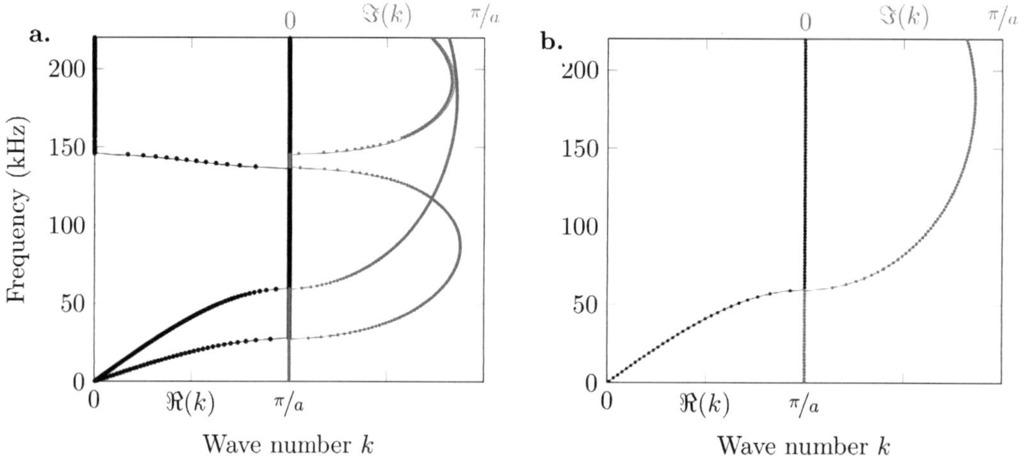

Figure 3. Dispersion curve of multilayer periodic structure with propagation at normal incidence without considering viscous losses. **a.** Transfer matrix method **b.** Bianco & Parodi method - black dots = real part of the wave number ; grey dots = imaginary part - On both figures, the analytical solution is represented by lines : longitudinal part (red), transverse part (blue)

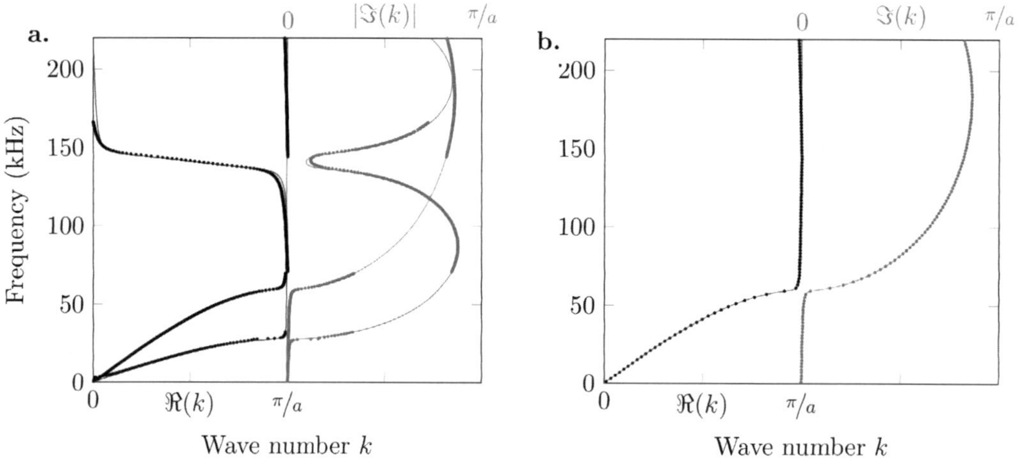

Figure 4. Dispersion curve of multilayer periodic structure with propagation at normal incidence considering viscous losses. **a.** Transfer matrix method **b.** Bianco & Parodi method - black dots = real part of the wave number ; grey dots = imaginary part - On both figures, the analytical solution is represented by lines : longitudinal part (red), transverse part (blue)

4.2. With losses

The purpose of this paper is to consider viscous losses in dispersion curves when considering dissipative materials. Both methods use the harmonic analysis. This computation in the *ATILA FEM* software allows complex values for material properties. An imaginary part for the wave number implies that these waves are evanescent. The same bi-layer structure used in the previous part is kept, but a very small amount of losses in epoxy, i.e. an imaginary part of the elasticity moduli equal to 2% of its real part, is considered.

Both methods are presented on figures 4a and 4b. Transfer matrix solutions are sometimes missing, due to the computing selection during resolution, but both methods give results in good agreement. One can see that curves are smoothed and stop-band boundaries are not strictly defined. Moreover, even in previous pass bands, an imaginary part is obtained : it means that the waves are partially evanescent due to viscous losses in the materials.

5. Conclusion

Two methods have been presented to retrieve the dispersion curve when materials have losses. The Bianco & Parodi method only gives information about the longitudinal propagation mode. To use this method, the slabs studied must be thick enough to allow a periodic behaviour. According to previous studies [15], slabs should be composed of 4 layers at least to obtain periodic effects. The transfer matrix method gives information about all propagative modes. However, the computation time is considerably heavier, especially in the 3D case. Moreover, the resolution is approximate and some results are missing due to the computing selection. Both methods give an imaginary part for the wave number, to get a readable representation. By taking into account losses, results show that previous pass bands have an imaginary part, and stop band boundaries are not strictly defined.

These tools can be used to optimize the dispersion of a periodic material in realistic conditions (i.e. with losses). By providing an unusual dispersion curve, simulations describe more accurate behaviour for periodic structures. Using both methods is also a good way to identify clearly which curve corresponds to a longitudinal or a transverse mode. Furthermore, whereas the transfer matrix method detects all propagative modes, Bianco & Parodi method is able to consider oblique propagation. Together, these methods appear to be a powerful way to put insight on physical phenomena encountered for both propagative and evanescent waves in periodic media.

By using the tools developed, it will allow to study new absorbing acoustic panels before experimental validation.

Acknowledgement

This study has been supported by DGA (France) via a PhD grant with DCNS Research.

References

[1] Gaunaurd G 1977 *Journal of the Acoustical Society of America* **62** 298–307
[2] Audoly C 1994 *Acta Acoustica* **2** 393–402
[3] Wei P J and Zhao Y P 2010 *Mechanics of Advanced Materials and Structures* **17** 383–392
[4] Hladky-Hennion A C and Decarpigny J N 1991 *Journal of the Acoustical Society of America* **90** 3356–3367
[5] ATILA, *Finite-Element Software Package for the analysis of 2D & 3D structures based on smart materials* Version 6.0.2 2010
[6] Langlet P 1993 *Finite element analysis of the propagation of acoustic waves in periodic materials* Ph.D. thesis Valenciennes University
[7] Langlet P, Hladky-Hennion A C and Decarpigny J N 1995 *Journal of the Acoustical Society of America* **98** 2792–2800
[8] Mead D J 1973 *Journal of Sound and Vibration* **27** 235–260
[9] Mead D J 1975 *Journal of Sound and Vibration* **40** 13–39
[10] Accorsi M L and Bennett M S 1991 *Journal of Sound and Vibration* **148** 279–292
[11] Bavencoffe M 2009 *Propagation of Lamb waves through a wave guide with a periodic interface* Ph.D. thesis Le Havre University
[12] Bianco B and Parodi M 1976 *Alta Frequenza* **45** 107 – 110
[13] Croenne C 2009 *Electromagnetic propagation and radiation control by metamaterials* Ph.D. thesis Lille 1 University
[14] Frickey D A 1994 *IEEE Transactions on Microwave Theory and Techniques* **42** 205 – 211
[15] Van Der Biest F 2005 *Multiple scattering and time reversal of acoustic waves in periodic and disordered media* Ph.D. thesis Paris 7 Diderot University

AUTHOR INDEX

Institute of Physics
Dirac House, Temple Back
Bristol BS1 6BE UK